Lalo Magni Riccardo Scattolini

Advanced and multivariable control

Società Editrice
ESCULAPIO

ISBN 978-88-9385-367-5

Corrected reprint 2020

Cover layout: Carlotta Lenzi

Printed in Italy by Digital Team - Fano (PU)

Società Editrice
ESCULAPIO

to Grazia, Federico, Cecilia, Carlo, Elda, Paolo, and all the family
(L. M.)

to Anna, Elena, and Carla
(R. S.)

Preface

The aim of this book is to present a number of topics usually not included in basic courses on System Theory and Automatic Control. In writing this textbook we have tried to combine a rigorous formulation with a style that emphasizes the main results and their possible application.
Chapter 1 is initially devoted to recall a number of concepts concerning the stability of continuous-time and discrete-time dynamic systems, and then to present the *Lyapunov theory*, a number of powerful results for the stability analysis for linear and nonlinear systems. Finally, it is shown how efficient nonlinear control techniques can be obtained with the Lyapunov theory in a quite simple way. The presentation is obviously non exhaustive, and the interested reader is referred to the books listed in the references and dealing with nonlinear control.
In Chapter 2, different norms of dynamic systems are introduced together with the definition of gain. These results will be useful in the following chapters to properly state a number of control problems where the selected norms of the closed-loop transfer functions will have to be minimized by the proper choice of the regulator. The chapter also introduces the popular *Small Gain Theorem*, a powerful tool for the analysis of feedback systems. The Small Gain Theorem will also be specialized to classes of feedback systems with static nonlinearities.
In Chapter 3 some basic techniques for the analysis and control design of Single-Input, Single-Output (*SISO*) systems are recalled, such as the stability criteria for closed-loop systems and the definition and role of the sensitivity functions. It is also shown how fundamental properties, such as stability, performance, and robustness can be stated in terms of the properties of the sensitivity functions. Finally, it is discussed how the control synthesis problem can be reformulated as the minimization of suitable system norms. This topic will be of interest to motivate the introduction, in the following chapters, of some synthesis techniques for Multi-Input, Multi-Output (*MIMO*) systems.
Chapter 4 presents the definition of poles and zeros of continuous-time *MIMO* systems, these definitions can also be used for discrete-time systems. Chapter 5 extends to the *MIMO* case many definitions and results for *SISO* systems, such as the manipulation rules of block diagrams, the frequency response, the stability of the feedback system, the static and dynamic performance.
Chapter 6 describes the *pole-placement*, or pole assignment, approach: a control synthesis technique which can be used both for *SISO* and for *MIMO* systems. The goal is to impose that the poles of closed-loop system lie in specified positions of the complex plane. First, algorithms are developed under the assumption that the system state is available (or measurable). Then, this assumption is removed by introducing suitable *state observers*. Both continuous-time and discrete-time are considered.
Chapter 7 is devoted to present the general *Hamilton-Jacobi-Bellmann* theory for the solution of optimal control problems. Nonlinear systems and generic cost

functions are considered. Then, in Chapter 8, this theory is specialized to linear continuous-time systems and quadratic cost functions. The celebrated *Linear Quadratic* (*LQ*) control is derived and its stability and robustness properties are presented. Chapter 9 summarizes the main assumptions and results concerning the *Kalman filter*, a well known observer which is optimal under suitable assumptions on the stochastic noises affecting the system. The main results of this chapter are proven by exploiting the duality between the Kalman filter and the *LQ* control.

The results reported in Chapters 8 and 9 are combined in Chapter 10 to state and solve the *Linear Quadratic Gaussian* (*LQG*) control problem for systems with stochastic disturbances. It is then shown that the *LQG* control can be seen as a specific instance of the more general H_2 problem, which consists in the minimization of a 2-norm of a suitable system. The final part of the chapter is devoted to give an hint on the general solution of the H_2 and H_∞ control synthesis techniques. The results presented in Chapters 7-10 for continuous-time systems are briefly extended to discrete-time systems in Chapter 11.

Finally, Chapter 12 presents the main ideas and results of *Model Predictive Control* (*MPC*), a family of control methods which has gained a wide popularity in view of its widespread diffusion in process control. The presentation of *MPC* is initially based on the results developed in the previous Chapters, to show how *MPC* can be viewed as a variant of finite horizon *LQ* control of discrete-time systems. The main characteristics and features of the most popular industrial *MPC* algorithms are discussed, and the stability properties of *MPC* are synthetically described.

Some useful mathematical notions are summarized in the Appendix.

Contents

1

Dynamic systems - stability

This chapter describes some fundamental stability results based on the Lyapunov theory. Continuous-time, linear and nonlinear dynamical systems are first considered. Then, the analysis is extended to discrete-time systems. The Lyapunov theory is finally used to compute stabilizing control laws for nonlinear systems.

1.1 Preliminaries

Consider the continuous-time, nonlinear, finite dimensional dynamical system described by

$$\dot{x}(t) = f\left(x(t), u(t)\right), \quad x(t_0) = x_0$$

where t is the time index, $x \in R^n$ is the state and $u \in R^m$ is the input. The system is *time-invariant*, since the function f does not explicitly depend on time. For this reason, it is always possible to assume that the initial time instant is $t_0 = 0$.

When there are no input variables, that is

$$\dot{x}(t) = f\left(x(t)\right)$$

the system is called *autonomous*.

The system is *linear* when the function f is linear in its state and input arguments, that is

$$\dot{x}(t) = Ax(t) + Bu(t)$$

where $A \in R^{n,n}$ and $B \in R^{n,m}$.

1.1.1 Equilibrium

The pair $(\bar{x}, \bar{u})$ is an equilibrium point if the following relation is fulfilled

$$f(\bar{x}, \bar{u}) = 0$$

that is if, starting from the initial state $\bar{x}$ with input $\bar{u}$, the system state indefinitely remains in $\bar{x}$.

For linear systems, the previous relation is equivalent to

$$A\bar{x} = -B\bar{u}$$

- if A does not have null eigenvalues, i.e. $\det A \neq 0$, for any constant input $\bar{u}$, the equilibrium state is
$$\bar{x} = -A^{-1}B\bar{u}$$

- if at least one eigenvalue of A is null, i.e. $\det A = 0$, the equation $A\bar{x} = -B\bar{u}$ can have infinite or no solutions, also depending on the value of $\bar{u}$.

Example 1.1.1. Let $\bar{u} = 0$ and

$$A = \begin{bmatrix} 0 & 1 \\ 0 & 0 \end{bmatrix}$$

The equilibrium condition

$$A\bar{x} = 0 \rightarrow \begin{bmatrix} 0 & 1 \\ 0 & 0 \end{bmatrix} \begin{bmatrix} \bar{x}_1 \\ \bar{x}_2 \end{bmatrix} = 0$$

has infinitely many solutions $\bar{x}_2 = 0$ for any $\bar{x}_1$.

On the other hand, assuming that

$$B\bar{u} = \begin{bmatrix} 0 \\ 1 \end{bmatrix}$$

the equation $A\bar{x} = -B\bar{u}$ does not have any solution.

1.1.2 Bifurcations

Consider the system
$$\dot{x}(t) = f\left(x(t), \alpha\right)$$

where α can be either a parameter or a constant input. Any variation of α can modify both the number and the characteristics of the equilibrium points. These values of α are called *bifurcations*.

Example 1.1.2. Consider the system

$$\dot{x}(t) = \alpha x(t) + x^3(t)$$

The equilibrium condition is
$$\bar{x}\left(\alpha + \bar{x}^2\right) = 0$$

that is

- $\bar{x} = 0$ if $\alpha \geq 0$
- $\bar{x} = 0$, $\bar{x} = \mp\sqrt{-\alpha}$ if $\alpha < 0$

In this case, $\alpha = 0$ is called a *pitchfork bifurcation* (see Figure 1.1).

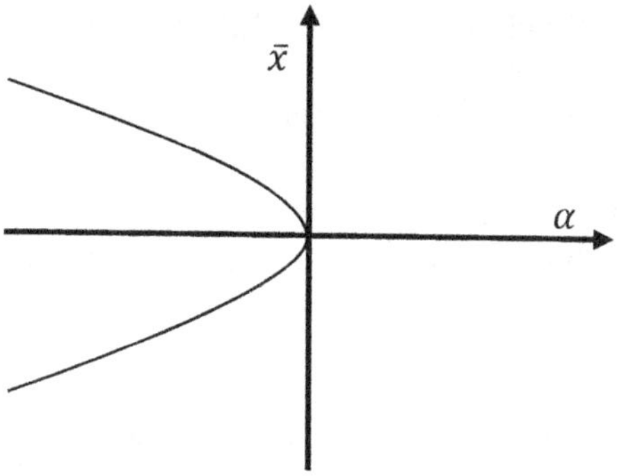

Figure 1.1: pitchfork bifurcation.

1.1.3 Equilibria and stability

Consider the system

$$\dot{x}(t) = f\left(x(t), u(t)\right)$$

the constant input $u(t) = \bar{u}$ and a corresponding equilibrium state $\bar{x}$.

Definition 1.1.1. The equilibrium $\bar{x}$ is *isolated* if there exists $\delta > 0$ such that there does not exist any other equilibrium in

$$B_\delta\left(\bar{x}, \delta\right) := \{x : \|x - \bar{x}\| \leq \delta\}$$

For second order systems, and considering the norm $\|x\|_2 = \sqrt{x_1^2 + x_2^2}$, this definition means that there are no other equilibria in the circle centered at $\bar{x}$ with radius δ, as shown in Figure 1.2.

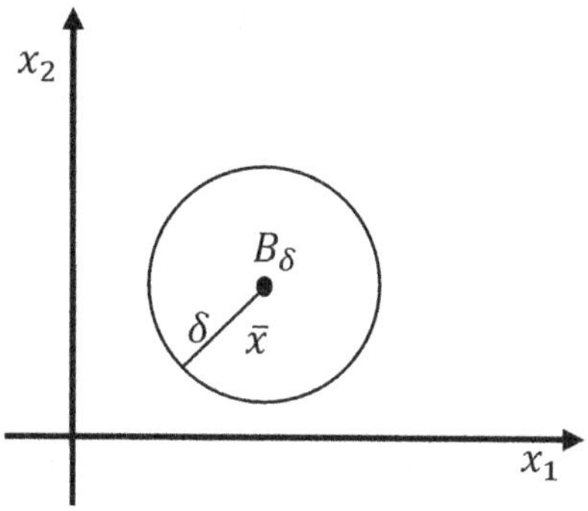

Figure 1.2: Isolated equilibrium: the only one equilibrium in B_δ is $\bar{x}$.

Note that, for any fixed $\bar{u}$, the system takes the form

$$\dot{x}(t) = f(x(t), \bar{u}) = \varphi(x(t))$$

Therefore, for simplicity and unless otherwise specified, in the following reference will be made to the autonomous system

$$\dot{x}(t) = \varphi(x(t))$$

and to the corresponding equilibrium specified by $\bar{x}$.

Definition 1.1.2. The equilibrium $\bar{x}$ is stable if for any $\varepsilon > 0$, there exists $\delta > 0$ such that, for all the initial states x_0 satisfying

$$\|x_0 - \bar{x}\| \leq \delta$$

it holds that

$$\|x(t) - \bar{x}\| \leq \varepsilon, \quad \forall t \geq 0$$

Definition 1.1.3. The equilibrium $\bar{x}$ is unstable if it is not stable.

Figure 1.3 graphically represents the definition of stability for a second order system. In particular, a possible state *trajectory* is shown, that is the projection on the state space (plane) of the state evolution $x(t)$, starting from x_0. The figure also shows the circles with radius δ and ε.

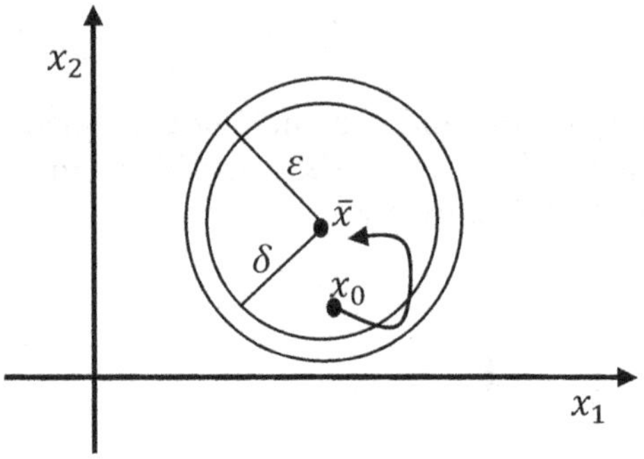

Figure 1.3: Stable equilibrium.

Definition 1.1.4. If $\bar{x}$ is a stable equilibrium and, in addition,

$$\lim_{t \to \infty} \|x(t) - \bar{x}\| = 0$$

then $\bar{x}$ is an *asymptotically stable* equilibrium.

Definition 1.1.5. The equilibrium $\bar{x}$ is *convergent* if there exists $\delta > 0$ such that, for all the initial states x_0 satisfying

$$\|x_0 - \bar{x}\| \leq \delta$$

it holds that

$$\lim_{t \to \infty} \|x(t) - \bar{x}\| = 0$$

Note that the definitions of asymptotic stability and convergence are different. In fact, convergence can be achieved even if the state trajectory goes indefinitely far from $\bar{x}$. However, if the equilibrium is both stable and convergent, then it is also asymptotically stable.

Remark. Asymptotic stability is a *local* property, since δ can be very small. The equilibrium $\bar{x}$ is *globally (asymptotically) stable* if it is asymptotically stable for any $x_0 \in R^n$.

Definition 1.1.6. The equilibrium $\bar{x}$ is *exponentially stable* if there exists $\delta > 0$ such that, for all the initial states x_0 satisfying

$$\|x_0 - \bar{x}\| \leq \delta$$

there exist two positive constants a and λ such that

$$\|x(t) - \bar{x}\| \leq a \|x_0 - \bar{x}\| e^{-\lambda t}, \quad \forall t \geq 0$$

It can be easily shown that the exponential stability implies the asymptotic stability.

Remark. Reference is often made to the stability of the origin $x = 0$. To this regard, note that, given the system

$$\dot{x}(t) = \varphi(x(t))$$

and the equilibrium $\bar{x}$ such that $\varphi(\bar{x}) = 0$, by means of the coordinates transformation $x(t) = \bar{x} + \delta x(t)$ it is possible to translate the equilibrium into the origin. In fact, the origin $\delta x = 0$ is an equilibrium of the system written as

$$\delta \dot{x}(t) = \varphi(\bar{x} + \delta x(t))$$

Remark. As already noted, in the case of non autonomous systems

$$\dot{x}(t) = f(x(t), u(t))$$

the equilibrium and its properties depend on the applied constant input.

Example 1.1.3. Consider the first order system

$$\dot{x}(t) = x(t)u(t)$$

The origin $\bar{x} = 0$ is an equilibrium both for $\bar{u} = 1$ and for $\bar{u} = -1$. Note however that when $\bar{u} = 1$ the system is $\dot{x}(t) = x(t)$ and the equilibrium $\bar{x} = 0$ is not stable, while for $\bar{u} = -1$ the system is $\dot{x}(t) = -x(t)$ and the equilibrium $\bar{x} = 0$ is globally asymptotically stable. The two cases are shown in Figure 1.4, where the arrows indicate if the state is increasing or decreasing. It is apparent that if $\dot{x}$ is positive the absolute value of the state grows, while if $\dot{x}$ is negative it decreases.

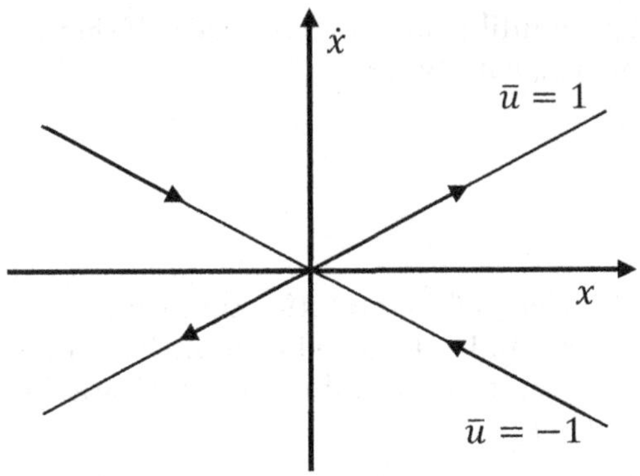

Figure 1.4: Stability of the equilibrium as function of the input.

1.1.4 Linear systems

For linear systems, stability is a property of the system itself, therefore it is a global property. The system is asymptotically stable if and only if all the eigenvalues of its dynamic matrix A have negative real part.

1.1.5 Nonlinear systems

Given the system

$$\dot{x}(t) = f\left(x(t), u(t)\right)$$

and the equilibrium defined by the pair $(\bar{x}, \bar{u})$, assuming that the function f is continuously differentiable with respect to its arguments, and letting

$$\begin{aligned} x(t) &= \bar{x} + \delta x(t) \\ u(t) &= \bar{u} + \delta u(t) \end{aligned}$$

it is possible to compute the linearized system

$$\delta\dot{x}(t) = A\delta x(t) + B\delta u(t)$$

where

$$A = \left.\frac{\partial f}{\partial x}\right|_{x=\bar{x}, u=\bar{u}}, \quad B = \left.\frac{\partial f}{\partial u}\right|_{x=\bar{x}, u=\bar{u}}$$

The following result holds:

- if all the eigenvalues of A have negative real part, then the equilibrium $(\bar{x}, \bar{u})$ is asymptotically stable;
- if at least one eigenvalue of A has positive real part, then the equilibrium $(\bar{x}, \bar{u})$ is unstable;

- if all the eigenvalues of A have negative or null real part, no conclusion can be drawn on the stability of the equilibrium from the analysis of the linearized system.

Note that, in any case, the previous result does not provide any information of the size of the stability region.

Example 1.1.4. Consider the system

$$\dot{x}(t) = -x(t) + x^2(t)$$

and the equilibrium $\bar{x} = 0$. The corresponding linearized model is

$$\delta\dot{x}(t) = -\delta x(t) + 2\bar{x}\delta x(t)$$

that is

$$\delta\dot{x}(t) = -\delta x(t)$$

Since the eigenvalue is -1, we can conclude that the equilibrium is asymptotically stable. In order to analyze the corresponding stability region, note that the analytic solution to the nonlinear system is

$$x(t) = \frac{e^{-t}x_0}{1 - x_0 + e^{-t}x_0}$$

The state evolution for a number of initial states x_0, is shown in Figure 1.5, it is apparent that the state asymptotically tends to zero for $x_0 < 1$.

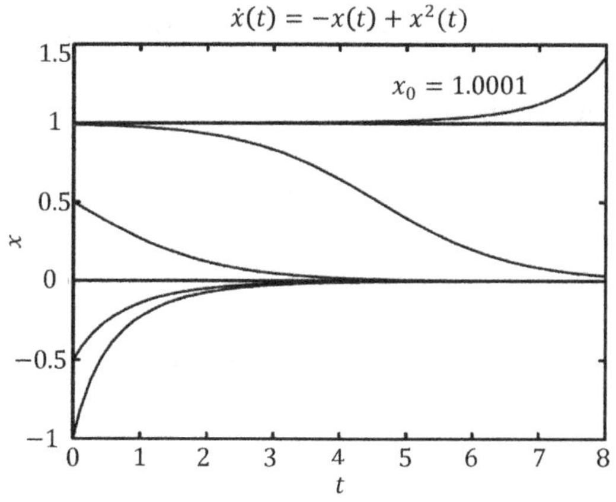

Figure 1.5: Example 1.1.4, state evolution.

The same conclusion can be drawn by analyzing the function $f(x)$, as in Figure 1.6. It is apparent that the equilibrium $\bar{x} = 0$ is asymptotically stable with region of attraction $(-\infty, 1)$. On the contrary, the equilibrium $\bar{x} = 1$ is unstable.

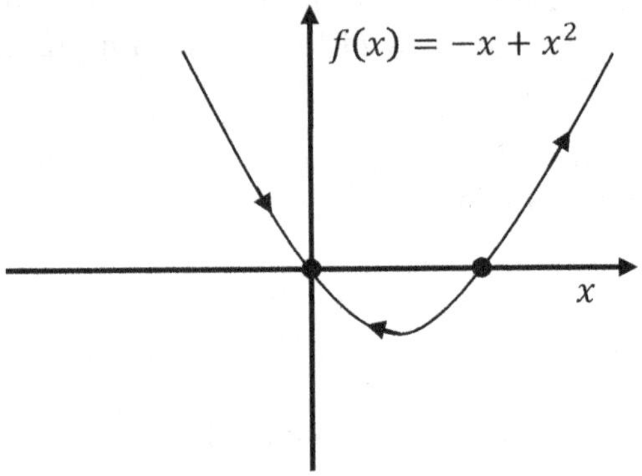

Figure 1.6: Example 1.1.4, equilibria and stability.

1.2 Second order systems - the phase plane

As shown in Example 1.1.4, for first order systems the stability of the equilibria can be analyzed by drawing $\dot{x}$ as a function of x. In the case of second order systems, it is possible to design the state trajectories in the so-called *phase plane* (x_1, x_2). In this way, one can study the behavior of the system for any initial condition. However, in nontrivial cases, this approach requires the use of proper computational procedures. Alternatively, in some cases, one of the following two analytical procedures can be followed.

Method (1). Analytically compute the state evolutions $x_1(t)$ and $x_2(t)$; then determine the function $x_2 = \phi(x_1)$.

Method (2). Write the system

$$\begin{aligned} \frac{dx_1(t)}{dt} &= \varphi_1(x_1(t), x_2(t)) \\ \frac{dx_2(t)}{dt} &= \varphi_2(x_1(t), x_2(t)) \end{aligned}$$

in the form

$$\frac{dx_2}{dx_1} = \frac{\varphi_2(x_1, x_2)}{\varphi_1(x_1, x_2)}$$

and analytically compute $x_2 = \phi(x_1)$.

1.2.1 Linear systems

For linear systems

$$\dot{x}(t) = Ax(t), \quad x = \begin{bmatrix} x_1 \\ x_2 \end{bmatrix}$$

some general results can be given.

Distinct eigenvalues

Assume that matrix A has distinct eigenvalues s_1 and s_2. Then there exists a state transformation $\hat{x} = Tx$ such that the transformed system

$$\dot{\hat{x}}(t) = TAT^{-1}\hat{x}(t) = A_d\hat{x}(t)$$

has a diagonal dynamic matrix

$$A_d = \begin{bmatrix} s_1 & 0 \\ 0 & s_2 \end{bmatrix}$$

Therefore

$$\hat{x}(t) = e^{A_d}\hat{x}(0) = \begin{bmatrix} e^{s_1 t} & 0 \\ 0 & e^{s_2 t} \end{bmatrix} \hat{x}(0)$$

and

$$x(t) = T^{-1} \begin{bmatrix} e^{s_1 t} & 0 \\ 0 & e^{s_2 t} \end{bmatrix} Tx(0)$$

It results that

$$x(t) = k_1 e^{s_1 t} + k_2 e^{s_2 t}$$

where k_1 and k_2 depend on the initial state x_0. The following cases must be considered.

1) $s_2 < s_1 < 0$

The origin is a *stable node* (see Figure 1.7.(a)).

If the initial state belongs to one of the eigenvectors r_1 and r_2, the whole state trajectory lies on the eigenvector itself. In other cases, the trajectory tends to the origin, where it is tangent to the eigenvector r_1 associated with s_1.

2) $s_2 > s_1 > 0$

The origin is an *unstable node* (see Figure 1.7.(b)).

This case is analogous to the previous one, but the trajectories diverge.

3) $s_2 < s_1 = 0$

By definition

$$Ar_1 = s_1 r_1$$

that is

$$Ar_1 = 0$$

Therefore, all the points in r_1 are equilibria. It is possible to prove that all the trajectories are parallel to r_2 and reach r_1 (see Figure 1.7.(c)).

4) $s_2 > s_1 = 0$

This case is analogous to the previous one, but the trajectories starting from the initial states $x_0 \notin r_1$ diverge (see Figure 1.7.(d)).

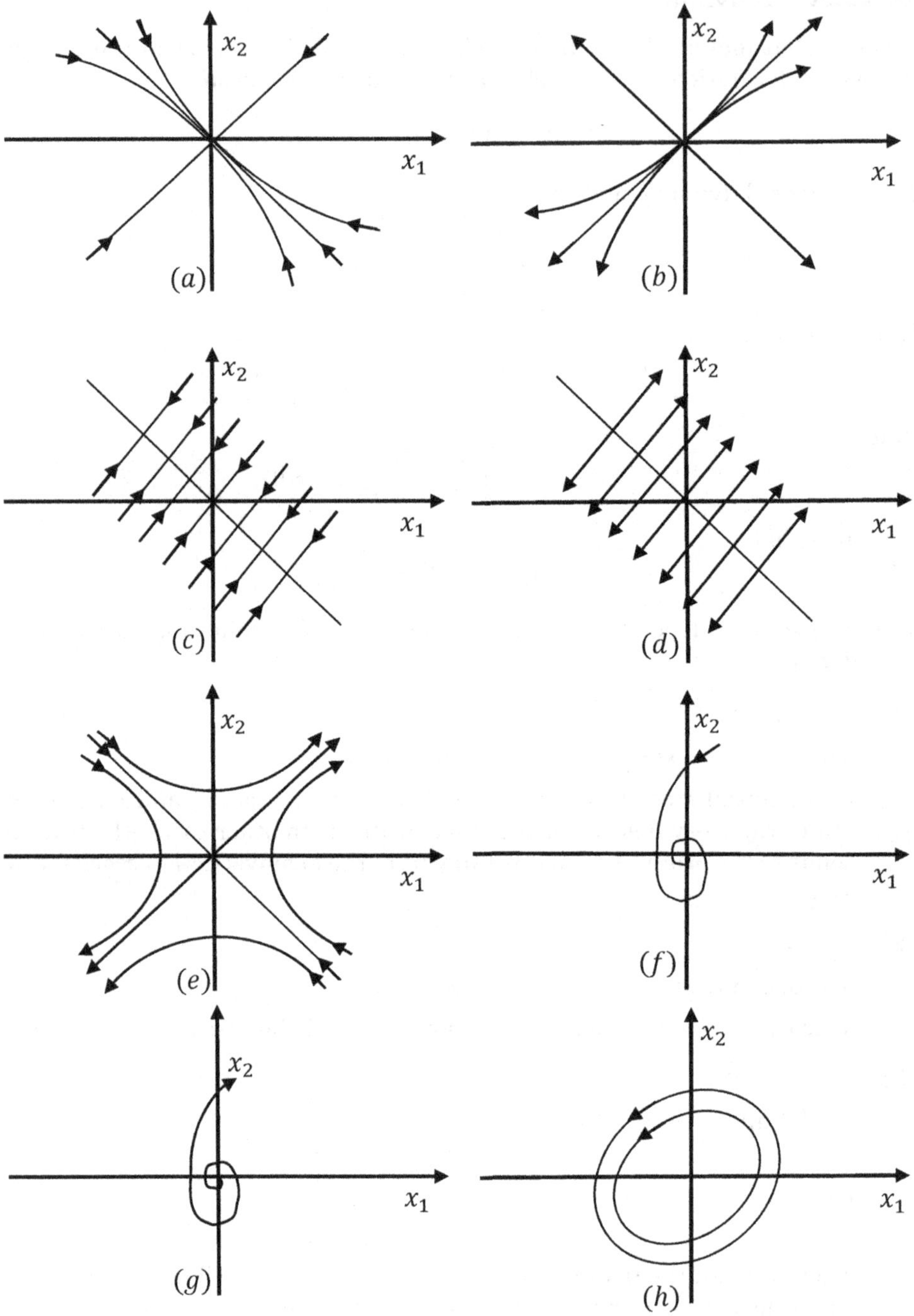

Figure 1.7: (a) $s_2 < s_1 < 0$; (b) $s_2 > s_1 > 0$; (c) $s_2 < s_1 = 0$; (d) $s_2 > s_1 = 0$; (e) $s_2 < 0, s_1 > 0$; (f) $s_{1,2} = a \pm jb, a < 0$; (g) $s_{1,2} = a \pm jb, a > 0$; (h) $s_{1,2} = a \pm jb, a = 0$.

5) $s_2 < 0$, $s_1 > 0$

The origin is a *saddle point* (see Figure 1.7.(e)).

6) $s_{1,2} = a \mp jb$

- if $a < 0$ the origin is a *stable focus* (see Figure 1.7.(f));
- if $a > 0$ the origin is a *unstable focus* (see Figure 1.7.(g));
- if $a = 0$ the origin is a *center* (see Figure 1.7.(h)).

Repeated eigenvalues

7) $s_2 = s_1 = s$, A diagonalizable.

If A is diagonalizable, there exists a transformation $\hat{x} = Tx$ such that $A_d = TAT^{-1} = sI$. In this case, any vector $\hat{x}$ is an eigenvector, since

$$A_d\hat{x} = sI\hat{x} = s\hat{x}$$

The origin is an *improper node*.

8) $s_2 = s_1 = s$, A non diagonalizable.

In this case, the system has an unique eigenvector and the trajectories, which depend on the sign of s, are depicted in Figure 1.8.

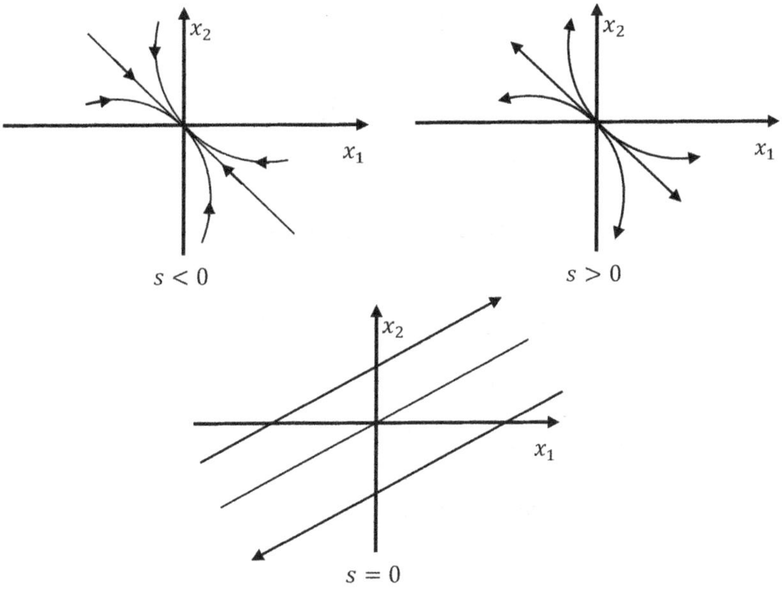

Figure 1.8: Non diagonalizable system.

1.2.2 Nonlinear systems

Consider the system

$$\dot{x}(t) = f\left(x(t), u(t)\right)$$

where f is continuous and differentiable with respect to its arguments. Then, in a neighborhood of the equilibrium $(\bar{x}, \bar{u})$, the phase plane of the nonlinear system coincides with that of the corresponding linearized system and its characteristics can be inferred by analyzing the eigenvalues of the linearized system, unless any eigenvalue of the linearized system has null real part.

Example 1.2.1. The pendulum shown in Figure 1.9 is described by the model

$$ML^2\ddot{\vartheta}(t) = -k\dot{\vartheta}(t) - MLg\sin\vartheta(t) + u(t), \quad M,\ L,\ k > 0$$

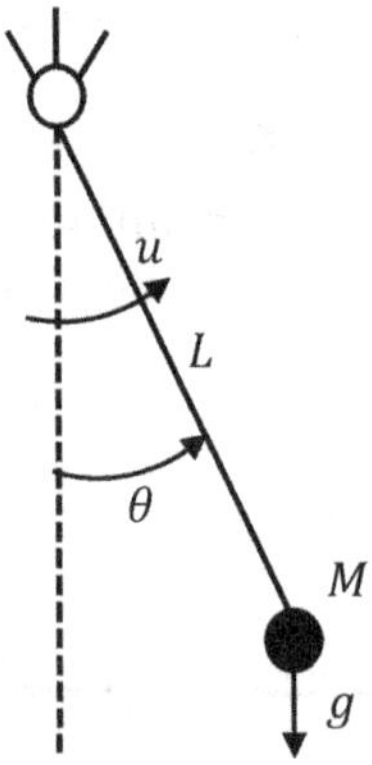

Figure 1.9: Pendulum.

Letting $x_1 = \vartheta$, $x_2 = \dot{\vartheta}$, the system can be written as

$$\begin{aligned} \dot{x}_1(t) &= x_2(t) \\ \dot{x}_2(t) &= -\frac{g}{L}\sin\left(x_1(t)\right) - \frac{k}{ML^2}x_2(t) + \frac{1}{ML^2}u(t) \end{aligned}$$

The equilibria corresponding to the input $\bar{u} = 0$ are

$$\begin{aligned} \bar{x}_2 &= 0 \\ \frac{g}{L}\sin\left(\bar{x}_1\right) &= 0 \quad \rightarrow \quad \bar{x}_1 = i\pi,\ i = 0, 1, ... \end{aligned}$$

It is apparent that the equilibria are asymptotically stable for $i = 0, 2, 4, ...$, while they are unstable for $i = 1, 3, 5, ...$

The linearized model is

$$\begin{aligned}\delta\dot{x}_1(t) &= \delta x_2(t)\\ \delta\dot{x}_2(t) &= -\frac{g}{L}\cos(\bar{x}_1)\,\delta x_1(t) - \frac{k}{ML^2}\delta x_2(t) - \frac{1}{ML^2}\delta u(t)\end{aligned}$$

and the corresponding dynamic matrix is

$$A = \begin{bmatrix} 0 & 1 \\ -\frac{g}{L}\cos(\bar{x}_1) & -\frac{k}{ML^2} \end{bmatrix}$$

Note that this matrix is in the *controllable canonical form* and, in order to be asymptotically stable, it must hold that $-\frac{g}{L}\cos(\bar{x}_1) < 0$, that is $\bar{x}_1 = 2\pi i$, $i = 0, 1, 2, ...$ On the contrary, for any $\bar{x}_1 = (2i+1)\pi$, $i = 0, 1, 2, ...$ the equilibria are unstable.

For example, given $M = 1$, $k = 1$, $L = 1$, $g = 9.8$, the eigenvalues are $s = -0.5 \mp j3$ ($\bar{x}_1 = 2\pi i$, stable focus) and $s_1 = 2.67$, $s_2 = -3.67$ ($\bar{x}_1 = (2i+1)\pi$, saddle point). The corresponding phase portrait is shown in Figure 1.10. Note that the region of attraction of the stable equilibria also depends on the initial speed $x_2(0)$.

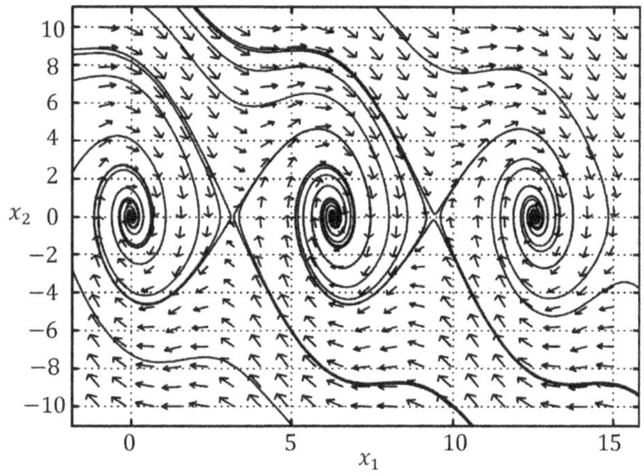

Figure 1.10: Pendulum - phase portrait ($M = L = k = 1$).

Example 1.2.2. Consider the van der Pol oscillator described by the model

$$m\ddot{x}(t) + 2c\left(x^2(t) - 1\right)\dot{x}(t) + kx(t) = 0, \quad m,\ c,\ k > 0$$

Letting $x_1 = x$, $x_2 = \dot{x}$, the following state space representation is obtained

$$\begin{aligned}\dot{x}_1(t) &= x_2(t)\\ \dot{x}_2(t) &= \frac{2c}{m}\left(1 - x_1^2(t)\right)x_2(t) - \frac{k}{m}x_1(t)\end{aligned}$$

The linearized model at the equilibrium $\bar{x}_1 = \bar{x}_2 = 0$, is

$$\begin{aligned}\delta\dot{x}_1(t) &= \delta x_2(t)\\ \delta\dot{x}_2(t) &= -\frac{k}{m}\delta x_1(t) + \frac{2c}{m}\delta x_2(t)\end{aligned}$$

The eigenvalues of the linearized system are

$$s = \frac{c}{m} \mp \sqrt{\frac{c^2}{m^2} - \frac{k}{m}}$$

The equilibrium is an unstable focus, as shown in Figure 1.11.

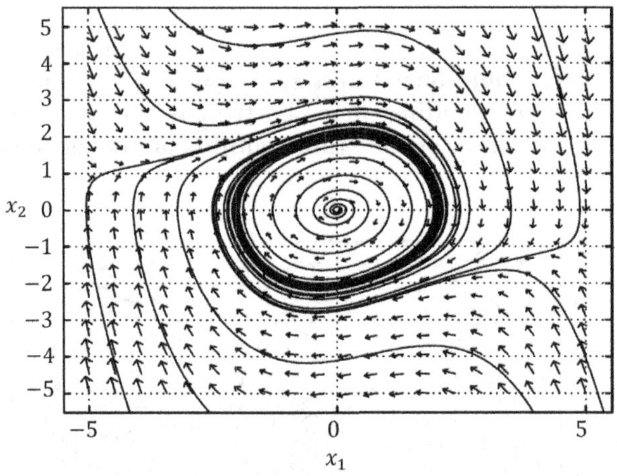

Figure 1.11: Van der Pol oscillator ($c = 0.1, m = 1, k = 1$).

From Figure 1.11 one can note that there exists a closed line, or *limit cycle*, where all the trajectories asymptotically converge.

Remark. According to the definition of Paragraph 1.1.3, the equilibrium of the previous example is unstable, even though there is a limit cycle and the trajectories do not depart indefinitely from the origin.

Remark. Also in the case of linear systems there can be cycles (centers) due to complex conjugate poles with null real part. However, these cycles depend on the initial condition and all the trajectory belongs to the cycle. For nonlinear systems, as shown in the previous example, the trajectory can reach a limit cycle starting from an initial condition not belonging to it.

Remark. For second order nonlinear systems there exist results, such as the Bendixon theorem, which allow one to determine analytically the existence of limit cycles.

1.3 The direct Lyapunov method

When the nonlinear system cannot be linearized at the equilibrium, the local analysis previously considered cannot be performed. For this reason, a general method to study the stability of the equilibrium is now described. The main idea underlying this approach is the following: if the total energy of the system is continuously dissipated, at least in a neighbor of the equilibrium, the system tends to move to the equilibrium itself, that turns out to be asymptotically stable.

Example 1.3.1. Consider the system shown in Figure 1.12, where the force associated with the spring is

$$k(x) = k_0 x + k_1 x^3$$

while the damper produces the dissipative force

$$h(x) = b\dot{x}\,|\dot{x}|$$

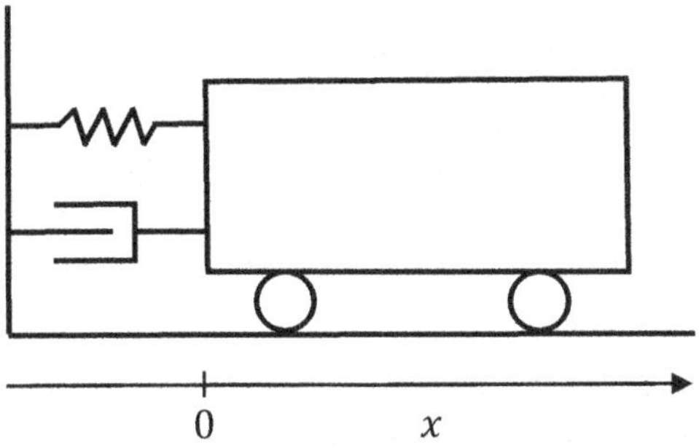

Figure 1.12: Cart.

The model of the system is (the dependence on time is neglected)

$$m\ddot{x} = -h(x) - k(x)$$

or

$$m\ddot{x} + b\dot{x}\,|\dot{x}| + k_0 x + k_1 x^3 = 0$$

The system is second order (states x and $\dot{x}$) with equilibrium $x = \dot{x} = 0$. However, the stability of the equilibrium cannot be established by analyzing the linearized model and, on the other hand, the explicit solution to the differential equation is not easy to compute.

The total energy of the system is $V(x, \dot{x}) = T(x, \dot{x}) + U(x, \dot{x})$, where the kinetic energy is

$$T(x, \dot{x}) = \frac{1}{2} m\dot{x}^2$$

and the potential energy is

$$\begin{aligned} U(x,\dot{x}) &= \int_0^x \left(k_0 x + k_1 x^3\right) dx \\ &= k_0 \frac{x^2}{2} + k_1 \frac{x^4}{4} \end{aligned}$$

It then follows that

$$V(x,\dot{x}) = \frac{1}{2} m\dot{x}^2 + k_0 \frac{x^2}{2} + k_1 \frac{x^4}{4}$$

and

$$\begin{aligned} \frac{dV}{dt} &= m\dot{x}\ddot{x} + k_0 x\dot{x} + k_1 x^3 \dot{x} \\ &= -b\dot{x}^2 \left|\dot{x}\right| = -b\left|\dot{x}^3\right| \end{aligned}$$

Therefore, one can conclude that $\dot{V} < 0$ for any $\dot{x} \neq 0$ and the energy is continuously dissipated until $\dot{V} = 0$, i.e. $\dot{x} = 0$.

Note that:

1. the energy function is positive for any $x \neq 0$ and $\dot{x} \neq 0$, while it is null only if the state is null ($x = \dot{x} = 0$);

2. its derivative $\dot{V}$ is negative for any $\dot{x} \neq 0$, while it is null only if $\dot{x}$ is null.

Definition 1.3.1. A *scalar* function $V(x)$, *continuous* with its first derivatives, is locally:

- *positive definite* in $\bar{x}$ if $V(\bar{x}) = 0$ and $V(x) > 0$ for any x belonging to an open neighbor of $\bar{x}$ (see Figure 1.13 where $\bar{x} = 0$);

- *semidefinite positive* in $\bar{x}$ if $V(\bar{x}) = 0$ and $V(x) \geq 0$ for any x belonging to an open neighbor of $\bar{x}$;

- *definite negative* in $\bar{x}$ if $V(\bar{x}) = 0$ and $V(x) < 0$ for any x belonging to a neighbor of $\bar{x}$;

- *semidefinite negative* in $\bar{x}$ if $V(\bar{x}) = 0$ and $V(x) \leq 0$ for any x belonging to an open neighbor of $\bar{x}$.

These definitions are *global* if the corresponding conditions are fulfilled for any x. Assume now that $V(x)$ is positive definite in $\bar{x}$ and $\lim_{\|x-\bar{x}\|\to\infty} V(x) = +\infty$, then $V(x)$ is *radially unbounded*.
Given a function $V(x)$ positive definite in $\bar{x}$, the values of x such that $V(x) = \bar{V}$, where $\bar{V}$ is a positive value, define a *level surface*, or a *level line* in the case of second order systems. With reference to this case, given $\bar{V} > 0$ sufficiently small, the corresponding level line is a closed curve that includes $\bar{x}$. In addition, given $\bar{V_2} > \bar{V_1} > 0$ the set of values of x such that $V(x) \leq V_1$ is contained into the set of values of x such that $V(x) \leq V_2$.

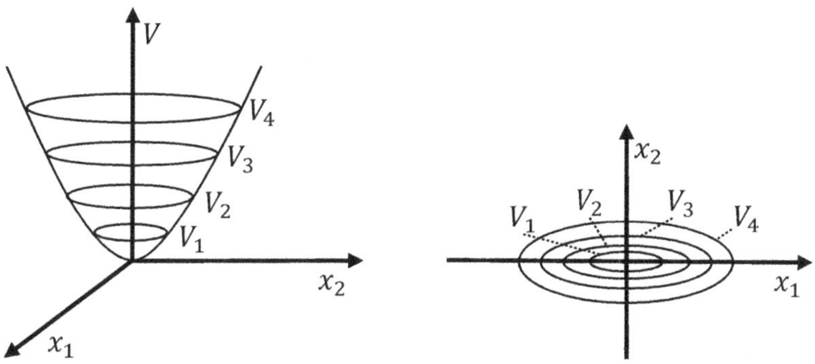

Figure 1.13: Positive definite function at the origin.

1.3.1 The Lyapunov theorem for nonlinear systems

Consider the system

$$\dot{x}(t) = \varphi\left(x(t)\right)$$

where $\varphi(\cdot)$ is continuous with its derivative. Letting $\bar{x}$ be an equilibrium, assume that $\varphi(\cdot))$ is locally Lipschitz in a neighbourhood D of $\bar{x}$, i.e. $\|\varphi(x) - \varphi(y)\| \leq L\|x - y\|, L > 0, x, y \in D$. Then, the following result holds.

Theorem 1.3.1. *If there exists a function $V(x)$, continuous with its derivative, positive definite in $\bar{x}$ and such that its derivative, along the state trajectories, $\dot{V}(x)$ is semidefinite negative in $\bar{x}$, i.e. such that for any x belonging to a neighbor of $\bar{x}$,*

$$\dot{V}(x) = \frac{dV}{dx}\frac{dx}{dt} = \frac{dV}{dx}\varphi\left(x\right) \leq 0$$

then $\bar{x}$ is a stable equilibrium.

Proof. Given $\varepsilon > 0$, consider the set $B_\varepsilon := \{x : \|x - \bar{x}\| < \varepsilon\}$. As shown in Figure 1.14 for second order systems, consider a closed level line $\bar{V}$ contained in B_ε. Define by δ the minimal distance of this curve from $\bar{x}$. Since by assumption $\dot{V} \leq 0$, for any initial state $x_0 \in B_\delta$, the state trajectory is fully contained in $\bar{V}$, i.e. it is inside B_ε.

A function $V(x)$ satisfying Theorem 1.3.3 is a *Lyapunov function* for the system.

The following results also hold, which allow one to check the asymptotic and global stability of the equilibrium.

Theorem 1.3.2. *If the assumptions of the previous Theorem hold and $\dot{V}$ is negative definite in $\bar{x}$, then $\bar{x}$ is an asymptotically stable equilibrium.*

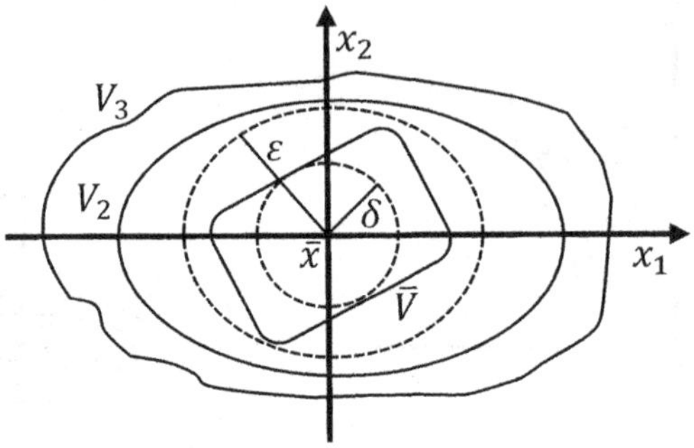

Figure 1.14: Lyapunov theorem.

Theorem 1.3.3. *If there exists a function $V(x)$, continuous with its derivative, globally positive definite in $\bar{x}$, radially unbounded, and such that its derivative, along the state trajectories, $\dot{V}(x)$ is globally definite negative in $\bar{x}$, then $\bar{x}$ is the unique globally asymptotically stable equilibrium of the system.*

Theorem 1.3.4 (Krasowski - La Salle Lemma). *If: (*a*) there exists a function $V(x)$ continuous with its derivative, positive definite in $\bar{x}$, such that $\dot{V}(x)$ is semidefinite negative in $\bar{x}$, (*b*) the set*

$$S := \left\{x : \dot{V}(x) = 0\right\}$$

does not contain perturbed (with respect to $\bar{x}$) trajectories compatible with the system, then $\bar{x}$ is an asymptotically stable equilibrium point.

Remark. The previous theorems are only sufficient conditions for the (asymptotic) stability of the equilibrium.

Theorem 1.3.5. *If there exists a function $V(x)$ continuous with its derivative, positive definite in $\bar{x}$ and such that $\dot{V}(x)$ is positive definite in $\bar{x}$, then $\bar{x}$ is an unstable equilibrium.*

Remark. In many cases, finding a proper $V(x)$ is not easy, even though it can be useful to associate it to the concept of energy, as in a previous example. Often one can set $V(x) = (x - \bar{x})' P (x - \bar{x})$, where P is a positive definite matrix, i.e. $v'Pv > 0$ for any $v \neq 0$.

Remark. Different Lyapunov functions can be found for the same system and equilibrium.

Example 1.3.2. Consider again Example 1.3.1, and recall that the state is $x_c = [x\ \dot{x}]'$, note that $V(x_c) > 0$ for $x_c \neq 0$ and $\dot{V}(x_c) = 0$ for $\dot{x} = 0$ and for any x. A priori, Theorem 1.3.1 allows one only to conclude that $x_c = 0$ is a stable, but not asymptotically stable, equilibrium. However, the set

$$S := \{x : \dot{x} = 0\}$$

coincides with the origin $x = 0$. In fact, $x \neq 0$ with $x \in S$, would lead to the model equation

$$m\ddot{x} = -k_0 x - k_1 x^3 \neq 0$$

that is $\ddot{x} \neq 0$ and the system could not remain in that value of x, that would not be an equilibrium. In view of the Krasowski - La Salle result, this argument allows one to conclude that $x = \dot{x} = 0$ is an asymptotically stable equilibrium.

1.3.2 The Lyapunov theorem for linear systems

The Lyapunov stability theory can be applied to the linear system

$$\dot{x}(t) = Ax(t) + Bu(t)$$

and the following result can be stated.

Theorem 1.3.6. *The necessary and sufficient condition for the asymptotic stability of the system is that, given any matrix $Q = Q' > 0$, there exists a matrix $P = P' > 0$ verifying the following Lyapunov equation*

$$A'P + PA = -Q$$

Proof. Note preliminarily that one can set $u = 0$.

Sufficiency Given any matrix $Q > 0$, assume that there exists P satisfying the Lyapunov equation and set $V(x) = x'Px$. It follows that

$$\begin{aligned} \dot{V} &= \dot{x}'Px + x'P\dot{x} = x'A'Px + x'PAx \\ &= x'(A'P + PA)x = -x'Qx < 0 \end{aligned}$$

Hence, V is a Lyapunov function for the system.

Necessity Consider $Q > 0$ and

$$P = P' = \int_0^\infty e^{A't}Qe^{At}dt > 0$$

It follows that

$$\begin{aligned} A'P + PA &= \int_0^\infty \left(A'e^{A't}Qe^{At} + e^{A't}Qe^{At}A\right) dt \\ &= \int_0^\infty \frac{d}{dt}\left(e^{A't}Qe^{At}\right) dt = \left[e^{A't}Qe^{At}\right]_0^\infty \end{aligned}$$

Being the system asymptotically stable, $e^{At} \overset{t\to\infty}{\to} 0$, therefore

$$\left[e^{A't}Qe^{At}\right]_0^\infty = -Q$$

that is

$$A'P + PA = -Q$$

In view of these results, a test for asymptotic stability consists of taking a generic matrix $Q > 0$, solving the Lyapunov equation with respect to P and verifying that $P > 0$. It is also possible to show that, given the matrix Q, if the system is asymptotically stable the solution to the Lyapunov equation is unique.

1.4 Discrete time systems

The previous results are now extended to invariant, discrete time systems described by

$$x(k+1) = f\,(x(k), u(k))$$

$$x(k+1) = f(x(k)) \qquad \text{autonomous systems}$$

$$x(k+1) = Ax(k) + Bu(k) \quad \text{linear systems}$$

1.4.1 Equilibrium

The pair $(\bar{x}, \bar{u})$ is an equilibrium if

$$\bar{x} = f(\bar{x}, \bar{u})$$

For linear systems this corresponds to

$$\bar{x} = A\bar{x} + B\bar{u}$$

that is

$$(I - A)\bar{x} = B\bar{u}$$

If the eigenvalues of A are different from 1, $\det(I - A) \neq 0$, and the equilibrium associated with a constant input $\bar{u}$ is given by

$$\bar{x} = (I - A)^{-1}B\bar{u}$$

1.4.2 Stability of the equilibrium

Definition 1.4.1. The equilibrium $\bar{x}$ is stable if for any $\varepsilon > 0$, there exists $\delta > 0$ such that, for all the initial states x_0 satisfying

$$\|x_0 - \bar{x}\| \leq \delta$$

one has

$$\|x(k) - \bar{x}\| \leq \varepsilon \quad , \quad \forall k \geq 0$$

The equilibrium is asymptotically stable if, in addition,

$$\lim_{k \to \infty} \|x(k) - \bar{x}\| = 0$$

Time invariant linear systems

For time invariant linear systems, stability is a property of the system. The necessary and sufficient condition for the asymptotic stability is that all the eigenvalues of A have modulus smaller than 1.

Nonlinear systems

Given the system

$$x(k+1) = f\left(x(k), u(k)\right)$$

and the equilibrium defined by the pair $(\bar{x}, \bar{u})$, letting f be continuously differentiable with respect to its arguments and defining

$$\begin{aligned} x(k) &= \bar{x} + \delta x(k) \\ u(k) &= \bar{u} + \delta u(k) \end{aligned}$$

the linearized system is

$$\delta x(k+1) = A\delta x(k) + B\delta u(k)$$

where

$$A = \left.\frac{\partial f}{\partial x}\right|_{x=\bar{x}, u=\bar{u}}, \qquad B = \left.\frac{\partial f}{\partial u}\right|_{x=\bar{x}, u=\bar{u}}$$

The following sufficient conditions hold:

- if all the eigenvalues of A have modulus smaller than 1, then the equilibrium $(\bar{x}, \bar{u})$ is asymptotically stable;
- if at least one eigenvalue of A has modulus greater than 1, then the equilibrium $(\bar{x}, \bar{u})$ is unstable;
- if all the eigenvalues of A have modulus smaller or equal to 1, nothing can be concluded on the stability properties of the equilibrium from the analysis of the linearized system.

1.4.3 The direct Lyapunov method

Consider the system

$$x(k+1) = \varphi(x(k))$$

with the equilibrium point $\bar{x}$. The following result holds.

Theorem 1.4.1. *If there exists a function $V(x)$ continuous and positive definite in $\bar{x}$ and such that*

$$\Delta V(x) = V(\varphi(x)) - V(x) \leq 0$$

in a neighbor of $\bar{x}$, that is if $\Delta V(x)$ is semidefinite negative in $\bar{x}$, then the equilibrium is stable. Moreover, if

$$\Delta V(x) < 0$$

in a neighbor of $\bar{x}$, then $\bar{x}$ is an asymptotically stable equilibrium.

Note that the condition on the derivative ($\dot{V}$) in the continuous time case has been transformed into a condition on the increment (ΔV) of V.

Theorem 1.4.2 (Krasowski - La Salle Lemma). *If there exists a function $V(x)$ positive definite in $\bar{x}$, with $\Delta V(x)$ negative semidefinite in $\bar{x}$ and the set*

$$S := \{x : \Delta V(x) = 0\}$$

does not contain perturbed (with respect to $\bar{x}$) trajectories compatible with the system, then $\bar{x}$ is an asymptotically stable equilibrium.

Linear systems

The direct application of the Lyapunov method to linear systems leads to the following result.

Theorem 1.4.3. *The necessary and sufficient condition for the asymptotic stability of the linear system*

$$x(k+1) = Ax(k)$$

is that for any matrix $Q = Q' > 0$ there exists a matrix $P = P' > 0$ solving the Lyapunov equation

$$A'PA - P = -Q$$

Proof. **Sufficiency** Given any $Q > 0$, assume that there exists $P > 0$ solving the Lyapunov equation. Now consider the Lyapunov function $V(x) = x'Px$. It follows that

$$\begin{aligned}
\Delta V(x) &= x'(k+1)Px(k+1) - x'(k)Px(k) \\
&= x'(k)A'PAx(k) - x'(k)Px(k) \\
&= x'(k)(A'PA - P)x(k) \\
&= -x'(k)Qx(k) < 0
\end{aligned}$$

Necessity Let A be asymptotically stable, take $Q > 0$ and

$$P = P' = \sum_{k=0}^{\infty} A'^k Q A^k > 0$$

Then

$$A'PA - P = A' \sum_{k=0}^{\infty} A'^k Q A^k A - \sum_{k=0}^{\infty} A'^k Q A^k = -Q$$

1.5 Control synthesis based on the Lyapunov theory

The Lyapunov theory can be used to design stabilizing control laws for nonlinear systems. This is shown first with an example, and then by introducing more systematic procedures.

Example 1.5.1. The model of a manipulator, described in the coordinates q, is

$$B(q)\ddot{q} + C(q,\dot{q})\dot{q} + F_v(q)\dot{q} + g(q) = \tau$$

where $B(q)\ddot{q}$ is the inertial term, $C(q,\dot{q})\dot{q}$ describes the centrifugal and Coriolis terms, $F_v(q)\dot{q}$ is a friction term, $g(q)$ is the gravitational term, and τ is the torque applied to the joints. The Lyapunov stability theory is now used to design a control law $\tau = \tau(q, q_d)$ with the aim to drive the manipulator to an equilibrium defined by the coordinates q_d. To this end, define the error vector $\tilde{q} = q_d - q$ and consider the Lyapunov function

$$V(\tilde{q},\dot{q}) = \frac{1}{2}\dot{q}' B(q)\dot{q} + \frac{1}{2}\tilde{q}' K_p\tilde{q}, \quad K_p > 0$$

where the first term describes the kinetic energy and the second one the potential energy. Note that V is a positive definite function with respect to $\tilde{q} = \dot{q} = 0$. With simple computations one obtains

$$\begin{aligned}
\dot{V} &= \dot{q}' B(q)\ddot{q} + \frac{1}{2}\dot{q}' \dot{B}(q)\dot{q} - \dot{q}' K_p\tilde{q} \\
&= \dot{q}' \left[-C(q,\dot{q})\dot{q} - F_v(q)\dot{q} - g(q) + \tau\right] + \frac{1}{2}\dot{q}' \dot{B}(q)\dot{q} - \dot{q}' K_p\tilde{q} \\
&= \frac{1}{2}\dot{q}' \left[\dot{B}(q) - 2C(q,\dot{q})\right]\dot{q} - \dot{q}' F_v(q)\dot{q} + \dot{q}' \left[\tau - g(q) - K_p\tilde{q}\right]
\end{aligned}$$

Looking at the structure of the matrices $B(q)$ and $C(q,\dot{q})$ it is possible to verify that $\dot{B}(q) - 2C(q,\dot{q}) = 0$. Consider now the following control law

$$\tau = g(q) + K_p\tilde{q} - K_d\dot{q}, \quad K_d > 0$$

with a proportional term on the error ($\tilde{q}$) and a derivative term on the output (q). From the previous expression one obtains

$$\dot{V} = -\dot{q}' F_v(q)\dot{q} - \dot{q}' K_d\dot{q}$$

It follows that $\dot{V} < 0$ when $\dot{q} \neq 0$, while $\dot{V} = 0$ for $\dot{q} = 0$ and for any $\tilde{q}$.

However, from the model equation and in view of the adopted control law it follows that

$$B(q)\ddot{q} + C(q,\dot{q})\dot{q} + F_v(q)\dot{q} = K_p\tilde{q} - K_d\dot{q}$$

Therefore, defining $S = \{\xi = \begin{bmatrix} \tilde{q} \\ \dot{q} \end{bmatrix} : \dot{q} = 0\}$, one has $K_p\tilde{q} = 0$, $\forall \xi \in S$ and, since $K_p > 0$, $\tilde{q} = 0$. In conclusion, $q = q_d$ and $\dot{q} = 0$ is the only equilibrium allowed for the system. Note that this is an asymptotically stable equilibrium.

1.5.1 Control Lyapunov functions

Consider the problem of designing a stabilizing control law $u = k(x)$ for the non-linear system

$$\dot{x}(t) = f(x(t)) + g(x(t))u(t), \quad x \in R^n, \quad u \in R^1$$

Assume to know a function $V(x) > 0$ continuous with its first derivative in $x = 0$ and such that

$$\inf_u \{V_x(x)f(x) + V_x(x)g(x)u\} \; < \; 0 \;, \; x \neq 0$$

where

$$V_x(x) \;=\; \frac{dV}{dx} = \left[\; \frac{dV}{dx_1} \quad \frac{dV}{dx_2} \quad \cdots \quad \frac{dV}{dx_n} \;\right]$$

This function V is called a *Control Lyapunov Function* (*CLF*) and allows one to define the following stabilizing control law

$$u(x) = \begin{cases} -\dfrac{V_x f(x) + \sqrt{(V_x f(x))^2 + (V_x g(x) g'(x) V_x')^2}}{V_x g(x) g'(x) V_x'} g'(x) V_x' & , \quad g'(x)V_x' \neq 0 \\ 0 & \quad g'(x)V_x' = 0 \end{cases}$$

This expression is named the *Sontag's* formula, and leads to

$$\dot{V}(x) = V_x[f(x) + g(x)u)] = \begin{cases} (a) & , \quad g'(x)V_x' \neq 0 \\ V_x f(x) < 0 & \quad g'(x)V_x' = 0 \end{cases}$$

where

$$\begin{aligned} (a) \;&=\; V_x f(x) - V_x g(x) \frac{V_x f(x) + \sqrt{(V_x f(x))^2 + (V_x g(x) g'(x) V_x')^2}}{V_x g(x) g'(x) V_x'} g'(x) V_x' \\ &=\; -\sqrt{(V_x f(x))^2 + (V_x g(x) g'(x) V_x')^2} < 0 \end{aligned}$$

Note that this result requires to know $V(x)$, that, in many cases, is a non trivial task.

1.5.2 The backstepping method

Consider the system

$$\begin{aligned} \dot{x}_1(t) \;&=\; f(x_1(t)) + g(x_1(t))x_2(t) \quad , \; x_1 \in R^n, \; x_2 \in R^1 \\ \dot{x}_2(t) \;&=\; u(t) \end{aligned}$$

where f and g are continuous and differentiable in a set $D \subset R^n$ and $f(0) = 0$. Now temporarily assume that x_2 is a free design variable, and to be able to stabilize the first equation of the system with the control law (the dependence on time is omitted to simplify the notation)

$$x_2 = \phi_1(x_1) \,, \; \phi_1(0) = 0$$

Moreover, with reference to the closed-loop system

$$\dot{x}_1 = f(x_1) + g(x_1)\phi_1(x_1)$$

assume to know a Lyapunov function $V_1(x_1)$ such that

$$\dot{V}_1(x_1) = \frac{dV_1}{dx_1}\left(f(x_1) + g(x_1)\phi_1(x_1)\right) = -W(x_1)$$

where $W(x_1) > 0$ for any $x_1 \in D$.
Now recall that the state variable x_2 cannot be freely chosen, and define the error $\eta = x_2 - \phi_1(x_1)$.
It follows that

$$\begin{aligned} \dot{\eta} &= \dot{x}_2 - \dot{\phi}_1(x_1) \\ &= u - \frac{\phi_1(x_1)}{dx_1}\dot{x}_1 = u - \frac{\phi_1(x_1)}{dx_1}[f(x_1) + g(x_1)(\eta + \phi_1(x_1))] \end{aligned}$$

Therefore, the original system can be written as

$$\begin{aligned} \dot{x}_1 &= f(x_1) + g(x_1)[\eta + \phi_1(x_1)] \\ \dot{\eta} &= u - \frac{\phi_1(x_1)}{dx_1}[f(x_1) + g(x_1)(\eta + \phi_1(x_1))] \end{aligned}$$

Now define a new input

$$v = u - \dot{\phi}_1(x_1)$$

Then $\dot{\eta} = v$ and the system can be given the form

$$\begin{aligned} \dot{x}_1 &= f(x_1) + g(x_1)\phi_1(x_1) + g(x_1)\eta \\ \dot{\eta} &= v \end{aligned}$$

where the origin is an asymptotically stable equilibrium of the first subsystem.
For the overall system consider now the Lyapunov function

$$V_2(x_1, \eta) = V_1(x_1) + \frac{1}{2}\eta^2 = V_1(x_1) + \frac{1}{2}(x_2 - \phi_1(x_1))^2$$

It follows that

$$\begin{aligned} \dot{V}_2(x_1, \eta) &= \frac{dV_1(x_1)}{dx_1}\dot{x}_1 + \eta\dot{\eta} \\ &= \frac{dV_1(x_1)}{dx_1}\left(f(x_1) + g(x_1)\phi_1(x_1) + g(x_1)\eta\right) + \eta v \\ &\leq -W(x_1) + \frac{dV_1(x_1)}{dx_1}g(x_1)\eta + \eta v \end{aligned}$$

Therefore, by imposing

$$v = -\frac{dV_1(x_1)}{dx_1}g(x_1) - k\eta \ , \ k > 0$$

it follows that

$$\dot{V}_2(x_1, \eta) = -W(x_1) - k\eta^2 < 0$$

and the equilibrium ($x_1 = 0, \eta = 0$) is asymptotically stable. Recalling that $\eta = x_2 - \phi_1(x_1)$ and $\phi_1(0) = 0$ it follows that ($x_1 = 0, x_2 = 0$) is an asymptotically stable equilibrium.

The overall control law, written as a function of the original coordinates, is

$$\begin{aligned} u &= v + \dot{\phi}_1(x_1) \\ &= -\frac{dV_1(x_1)}{dx_1} g(x_1) - k\eta + \dot{\phi}_1(x_1) \\ &= -\frac{dV_1(x_1)}{dx_1} g(x_1) - k\left(x_2 - \phi_1(x_1)\right) + \frac{d\phi_1(x_1)}{dx_1}\left(f(x_1) + g(x_1)x_2\right) \end{aligned}$$

This control law guarantees the asymptotic stability of the origin, with the associated Lyapunov function

$$\begin{aligned} V_2(x_1, x_2) &= V_1(x_1) + \frac{1}{2}\eta^2 \\ &= V_1(x_1) + \frac{1}{2}\left(x_2 - \phi_1(x_1)\right)^2 \end{aligned}$$

In conclusion, in order to compute the stabilizing control law $u = \kappa(x_1, x_2)$ it is only required to find the control law $x_2 = \phi_1(x_1)$ and the Lyapunov function $V_1(x_1)$ and then to apply the previous formula for u.

Example

Consider the system

$$\begin{aligned} \dot{x}_1 &= x_1^2 - x_1^3 + x_2 \\ \dot{x}_2 &= u \end{aligned}$$

and the control law

$$x_2 = \phi(x_1) = -x_1^2 - x_1$$

It follows that

$$\dot{x}_1 = -x_1^3 - x_1$$

Letting

$$V_1(x_1) = \frac{1}{2}x_1^2$$

one has

$$\dot{V}_1(x_1) = -x_1^4 - x_1^2 < 0$$

It is then possible to use the formula defining the control law $u = \kappa(x_1, x_2)$ with $k = 1$ by noting that

$$\begin{aligned} -\frac{dV_1(x_1)}{dx_1} g(x_1) &= -x_1 \\ -k\left(x_2 - \phi_1(x_1)\right) &= -(x_2 + x_1 + x_1^2) \\ \frac{d\phi_1(x_1)}{dx_1}\left(f(x_1) + g(x_1)x_2\right) &= -(2x_1 + 1)\left(x_1^2 - x_1^3 + x_2\right) \end{aligned}$$

Therefore

$$u = -x_1 - (x_2 + x_1 + x_1^2) - (2x_1 + 1)\left(x_1^2 - x_1^3 + x_2\right)$$

This control law globally stabilizes the origin of the controlled system, with Lyapunov function

$$\begin{aligned} V_2(x_1, \eta) &= V_1(x_1) + \frac{1}{2}\left(x_2 - \phi_1(x_1)\right)^2 \\ &= \frac{1}{2}x_1^2 + \frac{1}{2}\left(x_2 + x_1^2 + x_1\right)^2 \end{aligned}$$

1.5.3 A generalization

Consider now the system

$$\begin{aligned} \dot{x}_1(t) &= f_1(x_1(t)) + g_1(x_1(t))x_2(t) \quad , \; x_1 \in R^n, \; x_2 \in R^1 \\ \dot{x}_2(t) &= f_2(x_1(t), x_2(t)) + g_2(x_1(t), x_2(t))u(t) \end{aligned}$$

where it is also assumed that f_2 e g_2 are continuously differentiable functions and $g_2(x_1, x_2) \neq 0$ in the domain of interest. It is then possible to set

$$u = \frac{1}{g_2(x_1, x_2)}\left\{u_a - f_2(x_1, x_2)\right\}$$

where u_a is a fictitious input to be properly selected. It then follows that

$$\begin{aligned} \dot{x}_1(t) &= f_1(x_1) + g_1(x_1)x_2(t) \\ \dot{x}_2(t) &= u_a \end{aligned}$$

In view of the previous considerations, it is possible to conclude that, if $x_2 = \phi(x_1)$ stabilizes the origin of the first equation with the associated Lyapunov function $V_1(x_1)$, the required control law is

$$\begin{aligned} u &= \frac{1}{g_2(x_1, x_2)}\left\{\frac{d\phi_1(x_1)}{dx_1}\left(f_1(x_1) + g_1(x_1)x_2\right) - k\left(x_2 - \phi_1(x_1)\right)\right. \\ &- \left.\frac{dV_1(x_1)}{dx_1}g_1(x_1) - f_2(x_1, x_2)\right\} \end{aligned}$$

and the Lyapunov function associated with the feedback system is

$$V_2(x_1, x_2) = V_1(x_1) + \frac{1}{2}\left(x_2 - \phi_1(x_1)\right)^2$$

2
Norms, gain, and stability of feedback systems

In this Chapter, the definition of norms and gains of a system are given. Then the concept of input/output stability is introduced together with a criterion for the stability analysis of feedback systems, not necessarily composed only by the interconnection of linear systems.

2.1 Norms and gain of dynamical systems

Given a vector with m components

$$e = \begin{bmatrix} e_1 \\ e_2 \\ \vdots \\ e_m \end{bmatrix}$$

its 2-norm, or euclidean norm is

$$|e|_2 = \sqrt{e'e} = \sqrt{\sum_{i=1}^{m} e_i^2}$$

while its infinity norm is

$$|e|_\infty = \max_i |e_i|$$

Consider a time varying (vector) signal

$$e(t) = \begin{bmatrix} e_1(t) \\ e_2(t) \\ \vdots \\ e_m(t) \end{bmatrix}$$

for any fixed time t, it is possible to use the previous norms. Otherwise, the vector can be considered as a function of time and the following norms must be used

2-norm, or H_2 norm

$$\|e\|_2 = \sqrt{\int_{-\infty}^{+\infty} (e'(\tau)e(\tau))d\tau}$$

infinity norm, or H_∞ norm

$$\|e\|_\infty = \sup_t \left(\sup_i |e_i(t)| \right)$$

Consider a linear relation between two vectors y and u

$$y = Au$$

where A is a matrix of suitable dimensions. The norm of the "map" A is

$$|A| = \sup_{u\neq 0} \frac{|y|}{|u|} = \sup_{u\neq 0} \frac{|Au|}{|u|}$$

It follows that $|A|$ can be seen as the *gain* of the map. In particular, with the 2-norm one has

$$|y|_2^2 = y'y = u'A'Au \leq \lambda_{\max}(A'A)\,|u|_2^2$$

and, letting $\bar{\sigma}(A)$ be the maximum singular value of A,

$$\sup_{u\neq 0} \frac{|y|_2}{|u|_2} = \sqrt{\lambda_{\max}(A'A)} = \bar{\sigma}(A)$$

The definition of gain can be generalized to dynamical systems (non necessarily linear) with input u and output y, and described by

$$y(t) = S(u(t))$$

Let L_2 be the space of functions, null for $t < 0$, and whose absolute value raised to the 2nd power has finite integral, that is, if $u \in L_2$,

$$\|u\|_2 = \sqrt{\int_0^{+\infty} (u'(\tau)u(\tau))d\tau} < +\infty$$

Now assume that the system is $y(t) = S(u(t))$ and $u \in L_2$ (note that the output y does not need to belong to L_2). Then, the *gain* γ of S is defined as

$$\gamma = \|S\|_\infty = \sup_{u\in L_2} \frac{\|y\|_2}{\|u\|_2} = \sup_{u\in L_2} \frac{\|S(u)\|_2}{\|u\|_2}$$

and then

$$\|y\|_2 \leq \|S\|_\infty \|u\|_2 \quad , \quad \forall u \in L_2$$

Example
Consider an integrator

$$y(t) = \int_0^{+\infty} u(\tau)d\tau, \quad Y(s) = \frac{1}{s}U(s)$$

and the input $u \in L_2$ given by

$$u(t) = \begin{cases} 1 \ , & 0 < t < 1 \\ 0 \ , & t \geq 1 \end{cases}$$

It follows that

$$\|u\|_2 = \sqrt{\int_0^{+\infty} u'(\tau)u(\tau)d\tau} = 1$$

and

$$\|y\|_2 = +\infty$$

Therefore, the system has infinite gain (recall the modulus of the frequency response of $G(s) = 1/s$ for $\omega \to 0$).

Example
Consider a *Single Input, Single Output* ($SISO$) linear system described by the transfer function $G(s)$, or by the impulse response $g(\tau)$, such that $\int_{-\infty}^{+\infty} |g(\tau)|d\tau < +\infty$, and assume that

$$Y(s) = G(s)U(s)$$

Now assume that the input has Fourier transform $U(j\omega)$, so that

$$Y(j\omega) = G(j\omega)U(j\omega)$$

In view of the Parseval theorem one has

$$\|y\|_2^2 = \int_{-\infty}^{+\infty} y^2(\tau)d\tau = \frac{1}{2\pi}\int_{-\infty}^{+\infty} |Y(j\omega)|^2 d\omega = \frac{1}{2\pi}\int_{-\infty}^{+\infty} |G(j\omega)|^2 |U(j\omega)|^2 d\omega$$

If $|G(j\omega)| \leq K$ with $|G(j\omega)| = K$ for some ω, it follows that

$$\|y\|_2^2 \leq K^2\frac{1}{2\pi}\int_{-\infty}^{+\infty} |U(j\omega)|^2 d\omega = K^2 \int_{-\infty}^{+\infty} u^2(\tau)d\tau \leq K^2 \|u\|_2^2$$

and

$$\|G\|_\infty = \sup_\omega |G(j\omega)| = K$$

The gain is the supremum of the modulus of the frequency response of $G(s)$. This definition can be generalized to asymptotically stable *Multi Input, Multi Output* ($MIMO$) systems as follows

$$\gamma = \|G\|_\infty = \sup_\omega \bar{\sigma}\left(G(j\omega)\right)$$

Definition

A system $y = S(u)$ is input-output (I/O) stable if it has finite gain.

Note that, with reference to linear asymptotically stable systems, the term *gain* is often used with different meanings. In view of the previous considerations, we can refer to the following definitions

- *gain*, or *infinite-norm gain*: $\gamma = \|G\|_\infty$
- *gain at a given frequency* ω

$$\frac{\|Y(j\omega)\|_2}{\|U(j\omega)\|_2} = \frac{\|G(j\omega)U(j\omega)\|_2}{\|U(j\omega)\|_2}$$

  Therefore, for $SISO$ systems the gain at a given frequency ω is $|G(j\omega)|$, while for $MIMO$ systems the following relation holds, which associates the concept of gain to the minimum and maximum singular values of $G(j\omega)$

$$\underline{\sigma}(G(j\omega)) \leq \frac{\|G(j\omega)U(j\omega)\|_2}{\|U(j\omega)\|_2} \leq \bar{\sigma}(G(j\omega))$$

- *static gain*: gain at $\omega = 0$.

It is also possible to define a *2-norm gain* for strictly proper and asymptotically stable linear systems with transfer function $G(s)$ as follows.
SISO systems

$$\|G\|_2 = \sqrt{\frac{1}{2\pi}\int_{-\infty}^{+\infty} |G(j\omega)|^2 d\omega}$$

MIMO systems

$$\|G\|_2 = \sqrt{\frac{1}{2\pi}\int_{-\infty}^{+\infty} tr(G(j\omega)G'(-j\omega))d\omega}$$

This definition can be easily interpreted as follows. Letting $g(t)$ be the system impulse response, the 2-norm gain is

$$\|G\|_2 = \|g\|_2 = \sqrt{\int_0^{\infty} tr(g(\tau)'g(\tau))d\tau}$$

Then, given the asymptotically stable and strictly proper $SISO$ system with transfer function $G(s)$, input w and output z, i.e. $Z(s) = G(s)W(s)$, it follows that, if w is an impulse,

$$\|G\|_2 = \|z\|_2$$

2.2 Small gain theorem

Consider the feedback system shown in Figure 2.1, where S_1 and S_2 are I/O stable systems. The following *small gain theorem* can be proven

Theorem 2.2.1. *The feedback system is I/O stable if*

$$\|S_1\|_\infty \|S_2\|_\infty < 1$$

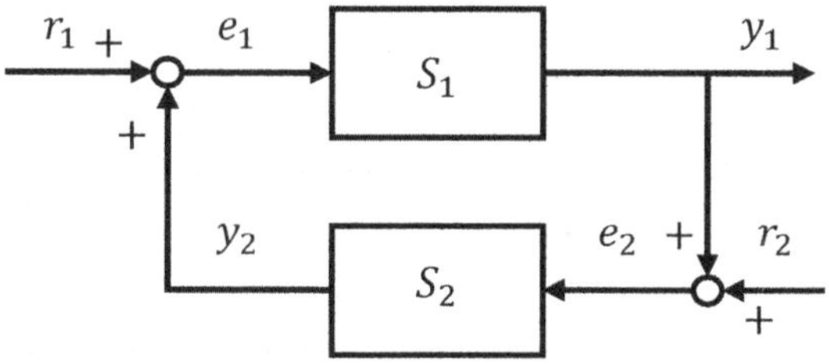

Figure 2.1: Feedback system.

To prove this result, note that

$$e_1 = r_1 + S_2(r_2 + y_1), \quad y_1 = S_1(e_1)$$

Therefore,

$$\|e_1\|_2 \leq \|r_1\|_2 + \|S_2\|_\infty (\|r_2\|_2 + \|S_1\|_\infty \|e_1\|_2)$$

It follows that

$$\|e_1\|_2 \leq \frac{\|r_1\|_2 + \|S_2\|_\infty \|r_2\|_2}{1 - \|S_1\|_\infty \|S_2\|_\infty}$$

and, with similar developments,

$$\|e_2\|_2 \leq \frac{\|r_2\|_2 + \|S_1\|_\infty \|r_1\|_2}{1 - \|S_1\|_\infty \|S_2\|_\infty}$$

Then, in view of the previous assumptions, the gain is finite.
It is also possible to prove that, if the two systems S_1 and S_2 are linear, the following less restrictive condition holds

$$\|S_1 S_2\|_\infty < 1$$

Considering again the linear case, for $SISO$ systems the previous result can be easily interpreted in terms of the Nyquist criterion. In fact, since by assumption S_1 and S_2 are asymptotically stable, the Nyquist diagram of the loop transfer function $S_1 S_2$ must not encircle the critical point -1. This condition is guaranteed by the previous condition, which implies that the diagram is entirely contained inside the circle with radius equal to one and centered at the origin.

2.3 The circle criterion

An interesting application of the small gain theorem allows one to study the I/O stability of the feedback system shown in Figure 2.2, where it is assumed that $G(s)$ is the transfer function of an asymptotically stable $SISO$ linear system and $f(\cdot)$ is a nonlinear, continuous function, uniquely defined for any input, such that $f(0) = 0$, and

$$k_1 e \leq f(e) \leq k_2 e$$

as shown in Figure 2.3(a).

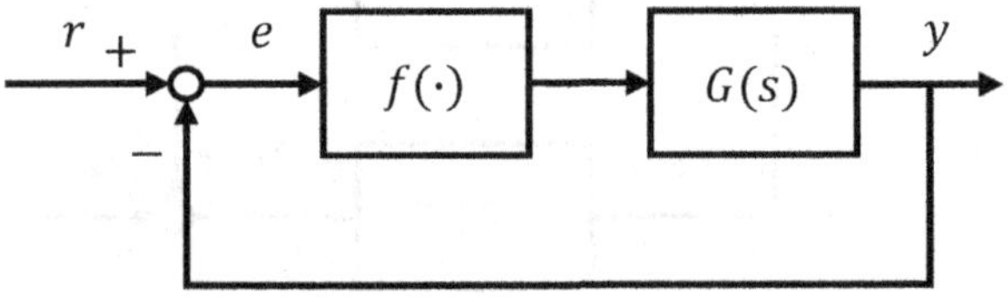

Figure 2.2: Feedback system.

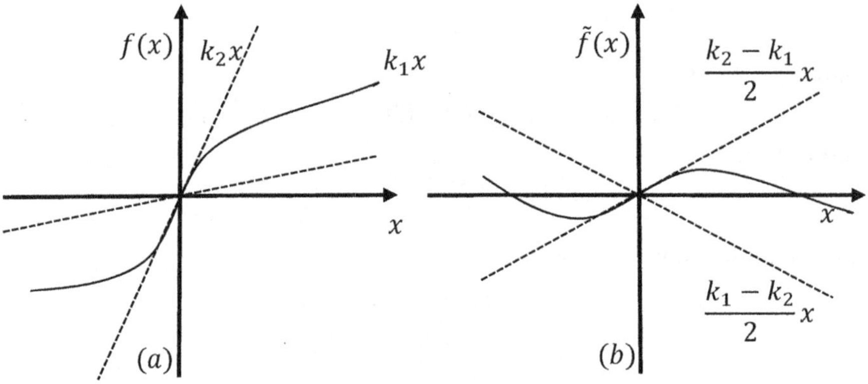

Figure 2.3: The considered nonlinear function (a) and function $\tilde{f}(\cdot)$ (b).

In view of the small gain theorem, the feedback system is I/O stable if

$$k_2 \sup_{\omega} |G(j\omega)| < 1$$

However, less stringent conditions can be obtained. Define

$$k = \frac{k_1 + k_2}{2}$$

and the function $\tilde{f}(\cdot)$ (see Figure 2.3(b)) such that

$$f(x) = kx + \tilde{f}(x)$$

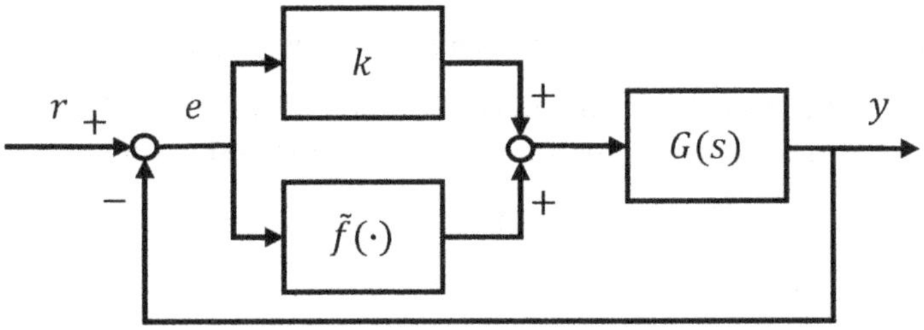

Figure 2.4: Reformulation of the control scheme.

Now draw the scheme as in Figure 2.4. By means of simple manipulations, it is possible to prove that

$$(1 + kG(s))e = r - G(s)\tilde{f}(e)$$

Letting

$$\tilde{G}(s) = \frac{G(s)}{1 + kG(s)}, \hat{G}(s) = \frac{1}{1 + kG(s)}$$

the scheme can be equivalently represented as in Figure 2.5.

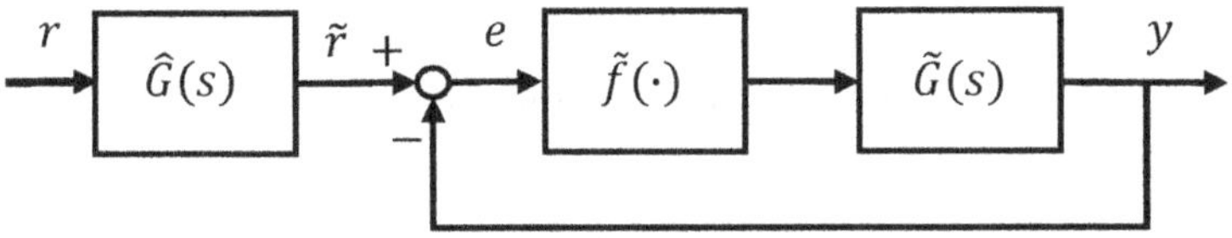

Figure 2.5: A different representation of the control scheme.

Define by

$$\gamma = \frac{k_2 - k_1}{2}$$

the gain of $\tilde{f}$ and assume that the transfer function $\tilde{G}(s)$ is asymptotically stable (this hypothesis will be proven in the following). In view of the small gain theorem, the feedback system is I/O stable if

$$\gamma \left|\tilde{G}(j\omega)\right| < 1, \quad \forall\omega$$

that is

$$\frac{1}{\left|\tilde{G}(j\omega)\right|} > \gamma \rightarrow \left|\frac{1}{G(j\omega)} + k\right| > \gamma$$

This is equivalent to require that, in the complex plane, the distance of the polar diagram of $\dfrac{1}{G(j\omega)}$ from $-k$ is greater than γ. Now transform any point z of the plane through the map $1/z$: the curve corresponding to $1/G(j\omega)$ maps into the one of $G(j\omega)$, while the previous limits are transformed as follows

$$-k-\gamma \rightarrow -\frac{1}{k_2}, \quad -k+\gamma \rightarrow -\frac{1}{k_1}$$

In conclusion, the I/O stability condition corresponds to require that the Nyquist diagram of $G(s)$ does not encircle the circle with diameter given by the segment $[-\dfrac{1}{k_1}, -\dfrac{1}{k_2}]$ and located on the x axis, as shown in Figure 2.6.

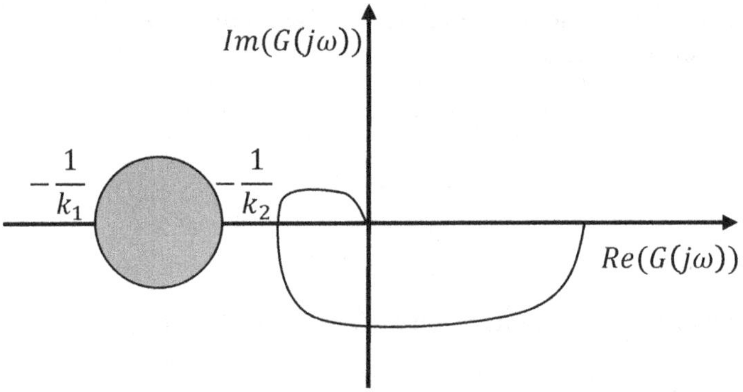

Figure 2.6: The circle criterion.

Note that the condition of the circle criterion implies that the diagram does not encircle $-\dfrac{1}{k}$. In view of the assumption of asymptotic stability of $G(s)$, this is equivalent to state that the system with transfer function $\tilde{G}(s)$ is asymptotically stable. In fact, this system can be equivalently represented as in Figure 2.7, and it is possible to apply the Nyquist criterion.

Remarks

1. The circle criterion can be extended to non asymptotically stable systems.
2. If k_1 tends to zero, the circle becomes a vertical line passing through $-1/k_2$, as shown in Figure 2.8. This case is of interest, since it allows one to consider nonlinearities typical of many systems, such as the saturation of the actuator or the dead zone reported in Figures 2.9.

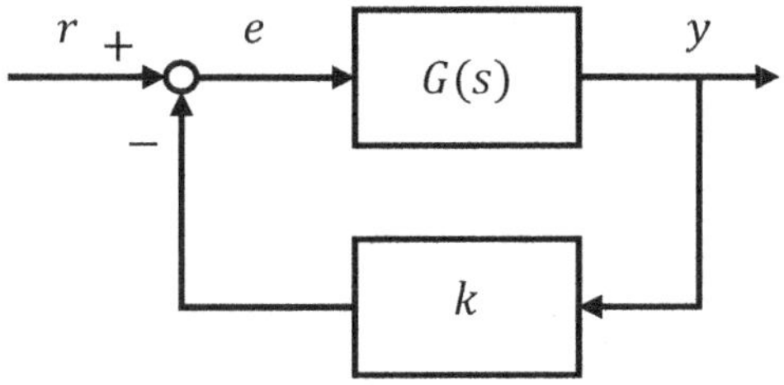

Figure 2.7: Equivalent scheme.

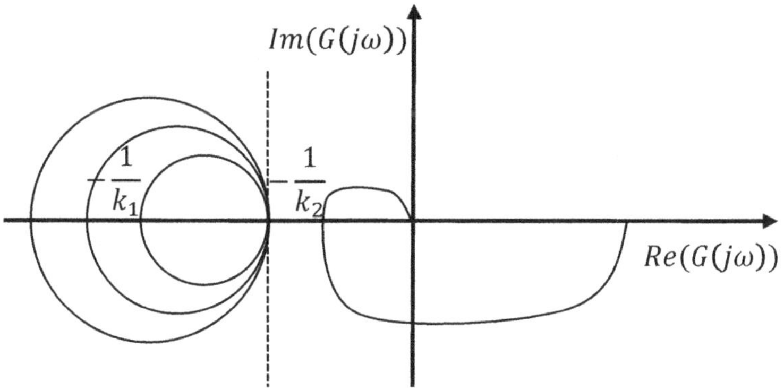

Figure 2.8: Circle criterion when $k_1 \longrightarrow 0$.

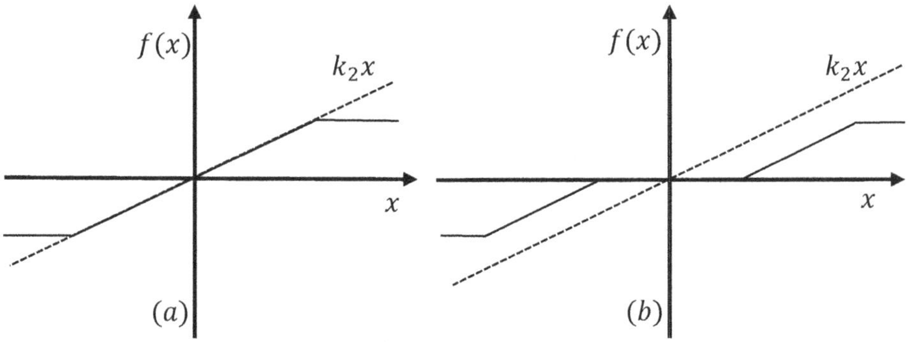

Figure 2.9: Saturation.

3 Control design for *SISO* systems - a brief overview

This chapter is devoted to recall some concepts concerning the synthesis of the controller for linear *SISO* systems. These concepts will be useful to introduce, in the following chapters, synthesis techniques for linear *MIMO* systems.
Consider the *SISO* feedback system shown in Figures 3.1 and 3.2.

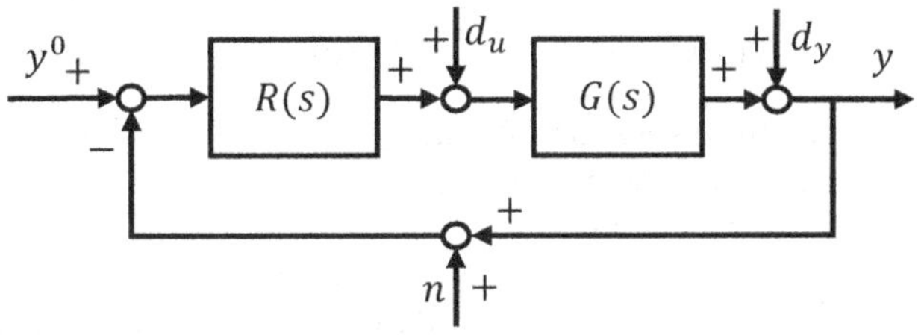

Figure 3.1: Control scheme.

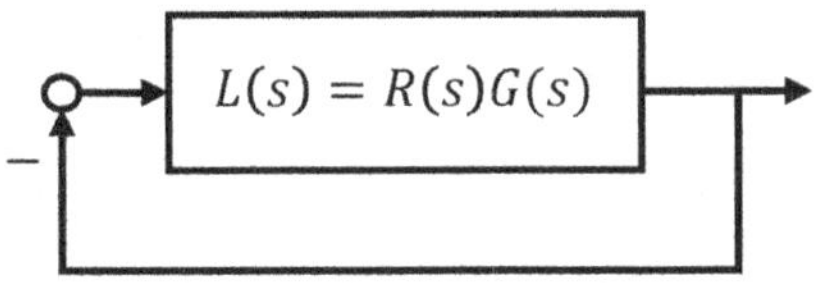

Figure 3.2: Control scheme with loop transfer function.

The relations among the exogenous signals and the main signals in the loop are:

$$
\begin{aligned}
Y(s) &= \frac{R(s)G(s)}{1+R(s)G(s)}(Y^o(s) - N(s)) + \frac{1}{1+R(s)G(s)}D_y(s) \\
&\quad + \frac{G(s)}{1+R(s)G(s)}D_u(s)
\end{aligned}
$$

$$U(s) = \frac{R(s)}{1+R(s)G(s)}(Y^o(s) - N(s) - D_y(s)) + \frac{1}{1+R(s)G(s)}D_u(s)$$

Define the following functions

- *Sensitivity* $S(s) = \dfrac{1}{1+R(s)G(s)}$
- *Complementary sensitivity* $T(s) = \dfrac{R(s)G(s)}{1+R(s)G(s)}$
- *Control sensitivity* $K(s) = \dfrac{R(s)}{1+R(s)G(s)} = S(s)R(s)$

It follows that

$$\begin{aligned} Y(s) &= T(s)(Y^o(s) - N(s)) + S(s)D_y(s) + S(s)G(s)D_u(s) \\ U(s) &= K(s)(Y^o(s) - N(s) - D_y(s)) + S(s)D_u(s) \end{aligned}$$

The main criteria to assess the stability of the closed-loop system based on the analysis of the open-loop transfer function $L(s)$ are the Nyquist criterion, the Bode criterion, and the root locus.
The *phase margin* φ_m and the *gain margin* g_m are two important parameters for the analysis of the properties of feedback systems, see Figure 3.3 for their definition. Both these parameters measure the distance of the Nyquist diagram of the loop transfer function $L(j\omega)$, at specified frequencies, from the critical point -1 (note however that in general they do not represent the minimum distance from -1).

The phase margin φ_m is useful to assess:

- stability (see the Bode criterion);
- robust stability (distance of the Nyquist diagram of $L(j\omega)$ from -1);
- performance: in many cases the complementary sensitivity function can be approximated as a second order transfer function, with poles characterized by damping $\varphi_m/100$. Hence, small values of the phase margin will lead to a high peak of the Bode diagram of $|T(j\omega)|$, that is to very oscillatory responses of the closed-loop system.

Moreover, the following approximate expressions hold:

$$|S(j\omega)| = \frac{1}{|1+L(j\omega)|} \simeq \begin{cases} \dfrac{1}{|L(j\omega)|}, & |L(j\omega)| \gg 1, \;\; \omega \ll \omega_c \\ 1, & |L(j\omega)| \ll 1, \;\; \omega \gg \omega_c \end{cases}$$

and

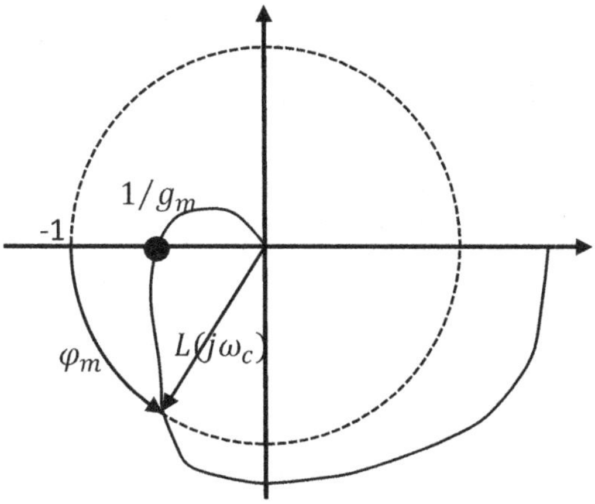

Figure 3.3: Phase and gain margin with the Nyquist diagram.

$$|T(j\omega)| = \frac{|L(j\omega)|}{|1+L(j\omega)|} \simeq \begin{cases} 1, & |L(j\omega)| \gg 1, \ \ \omega \ll \omega_c \\ |L(j\omega)|, & |L(j\omega)| \ll 1, \ \ \omega \gg \omega_c \end{cases}$$

where ω_c is the crossover (or critical) frequency.
Therefore, in general it is advisable to design a regulator so that:

- $|S|$ is "small" ($|L|$ "big") in the frequency band where the spectrum of the disturbance d_y has significant harmonic components (usually at low frequencies);
- $|T| \simeq 1$ ($|L|$ "big") in the frequency band where the spectrum of the reference signal has significant harmonic components (usually at low-medium frequencies);
- $|T|$ "small" ($|L|$ "small") in the frequency band where the spectrum of the measurement noise n has significant harmonic components (usually at high frequencies).

Recall that, if there are one or more integrators in $L(s)$, its modulus is high at low frequency (tending to infinity as ω tends to zero); this guarantees asymptotic null error for constant reference signals and disturbances d_y.

In view of the previous considerations, usually the regulator design must guarantee:

- $\varphi_m \geq \bar{\varphi}_m$;
- $g_m \geq \bar{g}_m$;

- $\omega_c \geq \bar{\omega}_c$, in order to guarantee satisfactory speed of response, tracking of reference signals, and attenuation of disturbances d_y. Note however that it is also advisable to set $\omega_c \leq \tilde{\omega}_c$ in order to attenuate the measurement noise n.

These requirements can be expressed in terms of a specified shape of the Bode diagram of $|L(j\omega)|$, as shown in Figure 3.4.

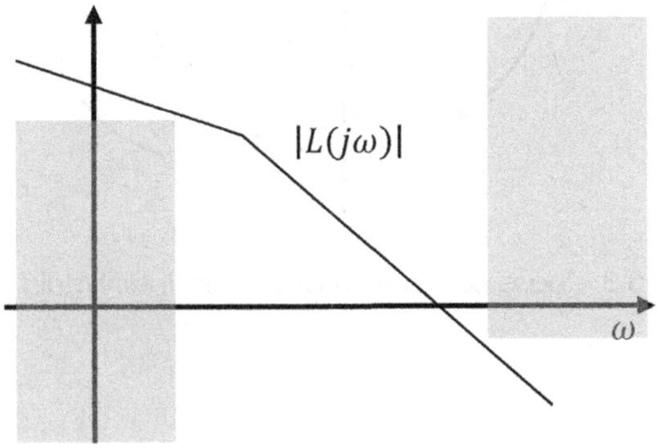

Figure 3.4: Bode diagram of the modulus of the loop transfer function.

3.1 Limits of performance

In the design of control systems, it is necessary to consider the following process characteristics, which can limit the achievable performance.

Time delay
Consider a system with delay τ between its input and its output. The system transfer function is $G(s)e^{-\tau s}$ where $G(s)$ is a rational term. The phase contribution of the delay is negative, and at the crossover frequency is $-\tau\omega_c$. If the goal is to guarantee a prescribed (positive) phase margin, the delay contribution must be limited, for example it could be required that $\tau\omega_c < \pi/6$. This corresponds to impose a limit to the crossover frequency, i.e. to the system's speed of response. For example, setting $\tau\omega_c < \pi/6$ one obtains

$$\omega_c < \frac{\pi}{6\tau} \simeq \frac{1}{2\tau}$$

Zeros with positive real part
The phase contribution of these zeros is negative, and their effect is similar to the one of time delay. Assume for example that the process transfer function is $G(s) = (1 - \tau s)G'(s)$, $\tau > 0$. The phase contribution of the zero placed in $1/\tau$ is

$$\varphi_z = -arctg(\tau\omega_c)$$

Therefore, if we want to impose $|\varphi_z| < \pi/6$ it is necessary that

$$\frac{\pi}{6} > arctg(\tau\omega_c) \rightarrow \omega_c < \frac{1}{\tau}tg(\frac{\pi}{6}) \simeq \frac{1}{2\tau}$$

Poles with positive real part
In this case, the closed-loop system must have a sufficiently high gain. This can be easily seen with the root locus technique: a high gain is required to guarantee that the position of the closed-loop poles, on the branches of the root locus leaving the unstable poles, is inside the stability region, i.e. sufficiently far from the open-loop poles. Consider for example a control system with transfer function $G(s) = \rho/(s-1)$. The root locus (positive gain) is shown in Figure 3.5.

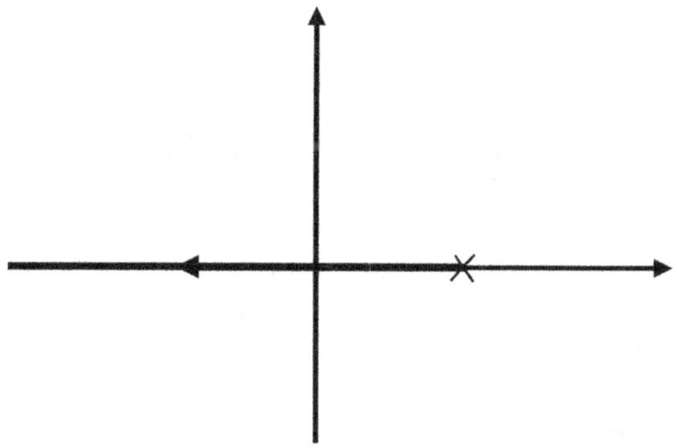

Figure 3.5: Root locus.

Limits of the sensitivity function
It is possible to prove the following theorem (Bode formula).

Theorem 3.1.1. *Consider an asymptotically stable closed-loop system, assume that the loop transfer function $L(s)$ does not have poles with positive real part, and has a relative degree greater or equal to 2. Then*

$$\int_0^{\infty} |S(j\omega)|_{dB}\, d\omega = 0$$

This result, which could be generalized also to unstable systems, means that the area of sensitivity reduction ($|S(j\omega)|$ below the $0dB$ axis), equals the area of sensitivity increase ($|S(j\omega)|$ above the $0dB$ axis). This is the so-called *waterbed effect*, and is shown in Figure 3.6.

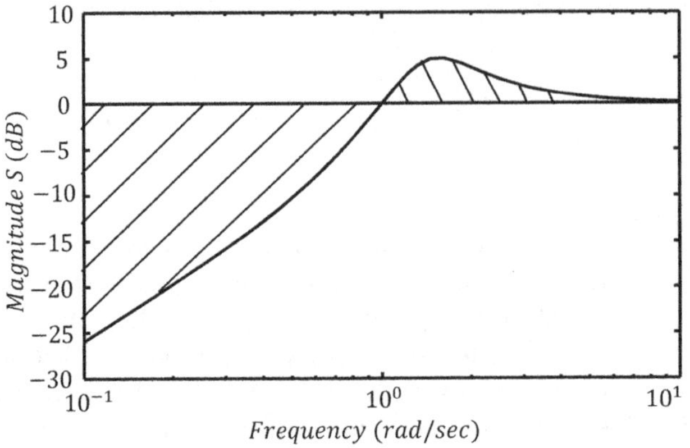

Figure 3.6: Sensitivity function and *waterbed effect.*

3.2 Sensitivity functions and design parameters

The approximations, previously described, of the sensitivity functions $|S(j\omega)|$ and $|T(j\omega)|$ can be very poor in the neighbors of the crossover frequency ω_c, where both $|S(j\omega)|$ and $|T(j\omega)|$ can have a peak which must be limited to guarantee suitable performance (for example, to avoid amplification of the disturbance d_y or oscillating responses due to step variations of the reference signal).

Now recall that, given a *SISO* system with a proper and asymptotically stable transfer function $\Gamma(s)$, its infinity norm H_∞ is denoted by $\|\Gamma\|_\infty$. Then, the peaks of $|S(j\omega)|$ and $|T(j\omega)|$ are

$$M_S = \sup_\omega |S(j\omega)| = \|S\|_\infty$$

and

$$M_T = \sup_\omega |T(j\omega)| = \|T\|_\infty$$

Note also that M_s^{-1} is the minimum distance of the Nyquist diagram of $L(j\omega)$ from -1 (see Figure 3.7).

Moreover, since

$$S(s) + T(s) = 1$$

it holds that

$$||S(j\omega)| - |T(j\omega)|| \leq |S(j\omega) + T(j\omega)| = 1, \quad \forall\omega$$

and, in particular, M_S and M_T differ at most by 1.

In view of these considerations, in the regulator design one can consider the following specifications

$$\begin{cases} M_s \leq \bar{M}_s \\ M_T \leq \bar{M}_T \end{cases}$$

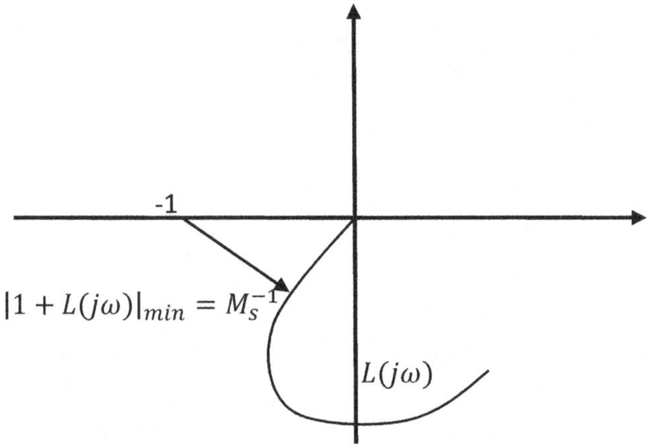

Figure 3.7: Minimum distance from -1.

Typical choices are $\bar{M}_s = 2$ $(6dB)$ and $\bar{M}_T = 1.5$ $(3.5dB)$.

The relations between these parameters and the phase margin φ_m and gain margin g_m can be further analyzed.

Define by ω_π the frequency where the phase of $L(j\omega)$ is $-\pi$ and assume $g_m > 1$. In view of the definition of the gain margin, one has $L(j\omega_\pi) = -1/g_m$, which implies

$$T(j\omega_\pi) = \frac{-1/g_m}{1 - 1/g_m}$$

Therefore

$$M_T \geq |T(j\omega_\pi)| = \frac{1}{g_m - 1}$$

and

$$g_m \geq 1 + \frac{1}{M_T}$$

Analogously

$$M_S \geq |S(j\omega_\pi)| = \frac{1}{1 - 1/g_m}$$

that is

$$g_m \geq \frac{M_S}{M_S - 1}$$

Now consider the phase margin φ_m, and note that

$$\frac{1}{M_S} \leq |1 + L(j\omega_c)|$$

and, for $\omega = \omega_c$ (see Figure 3.8)

$$\frac{1}{M_S} \leq |1 + L(j\omega_c)| = 2 \sin \varphi_m/2$$

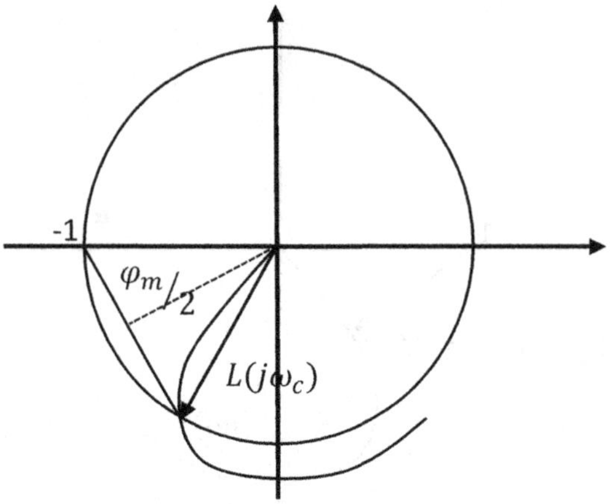

Figure 3.8: Sensitivity and phase margin.

It then follows that

$$\varphi_m \geq 2\arcsin\left(\frac{1}{2M_S}\right) \geq \frac{1}{M_S}$$

Note also that

$$|T(j\omega_c)| = \frac{|L(j\omega_c)|}{|1+L(j\omega_c)|} = \frac{1}{|1+L(j\omega_c)|} = |S(j\omega_c)|$$

Therefore, it holds that

$$\varphi_m \geq 2\arcsin\left(\frac{1}{2M_T}\right) \geq \frac{1}{M_T}$$

3.3 Sensitivity and parametric uncertainties

Assume that the process transfer function is $G(s,\vartheta)$, where ϑ is a set of parameters with known nominal value $\bar{\vartheta}_n$. In order to study the sensitivity of the closed-loop system, designed in the nominal case, to parameters' variations $\vartheta = \bar{\vartheta}_n + \Delta\vartheta$, set

$$G(s,\vartheta) \simeq G(s,\bar{\vartheta}_n) + \left.\frac{\partial G(s,\vartheta)}{\partial\vartheta}\right|_{\vartheta=\bar{\vartheta}_n} \Delta\vartheta$$

Define

$$\Delta G(s,\vartheta,\bar{\vartheta}_n) = \left.\frac{\partial G(s,\vartheta)}{\partial\vartheta}\right|_{\vartheta=\bar{\vartheta}_n} \Delta\vartheta$$

and

$$E_G(s,\vartheta,\bar{\vartheta}_n) = \frac{\Delta G(s,\vartheta,\bar{\vartheta}_n)}{G(s,\bar{\vartheta}_n)}$$

The term $E_G(s,\vartheta,\bar{\vartheta}_n)$ is the relative variation of the open-loop transfer function and, for $s = j\omega$, $E_G(j\omega,\vartheta,\bar{\vartheta}_n)$ represents the effect of the parametric uncertainty on the frequency response.

Now consider the closed-loop control system with complementary sensitivity function

$$T(s,\vartheta) = \frac{G(s,\vartheta)R(s)}{1+G(s,\vartheta)R(s)}$$

It holds that

$$\begin{aligned} T(s,\vartheta) &= T(s,\bar{\vartheta}_n) + \left.\frac{\partial T(s,\vartheta)}{\partial\vartheta}\right|_{\vartheta=\bar{\vartheta}_n} \Delta\vartheta \\ &= T(s,\bar{\vartheta}_n) + \frac{R(s)}{(1+G(s,\bar{\vartheta}_n)R(s))^2} \left.\frac{\partial G(s,\vartheta)}{\partial\vartheta}\right|_{\vartheta=\bar{\vartheta}_n} \Delta\vartheta \end{aligned}$$

Therefore

$$\Delta T(s,\vartheta,\bar{\vartheta}_n) = T(s,\vartheta) - T(s,\bar{\vartheta}_n) = \frac{R(s)}{(1+G(s,\bar{\vartheta}_n)R(s))^2} \left.\frac{\partial G(s,\vartheta)}{\partial\vartheta}\right|_{\vartheta=\bar{\vartheta}_n} \Delta\vartheta$$

and

$$\begin{aligned} E_T(s,\vartheta,\bar{\vartheta}_n) &= \frac{\Delta T(s,\vartheta,\bar{\vartheta}_n)}{T(s,\bar{\vartheta}_n)} \\ &\simeq \frac{1}{G(s,\bar{\vartheta}_n)(1+G(s,\bar{\vartheta}_n)R(s))} \left.\frac{\partial G(s,\vartheta)}{\partial\vartheta}\right|_{\vartheta=\bar{\vartheta}_n} \Delta\vartheta \\ &= \frac{\Delta G(s,\vartheta,\bar{\vartheta}_n)}{G(s,\bar{\vartheta}_n)} \frac{1}{1+G(s,\bar{\vartheta}_n)R(s)} = S(s,\bar{\vartheta}_n)E_G(s,\vartheta,\bar{\vartheta}_n) \end{aligned}$$

Finally,

$$S(j\omega,\bar{\vartheta}_n) = \frac{E_T(j\omega,\vartheta,\bar{\vartheta}_n)}{E_G(j\omega,\vartheta,\bar{\vartheta}_n)}$$

It is then possible to conclude that the nominal sensitivity function is the ratio between the closed-loop and open-loop effects of the uncertainty. This shows the attenuation provided by the closed-loop system on the uncertainty effects in the frequency band where $|S(j\omega,\bar{\vartheta}_n)| \ll 1$.

3.4 Robustness with respect to additive uncertainty

Consider the feedback control system of Figure 3.2 and assume that the system is described by

$$G(s) = \bar{G}(s) + \Delta G_a(s)$$

where $\bar{G}(s)$ is the nominal transfer function. Moreover, assume that

- the closed-loop nominal system is asymptotically stable;

- $\Delta G_a(s)$ is an asymptotically stable transfer function. Letting $P_{\bar{G}}$ and P_G be the number of poles of $\bar{G}(s)$ and $G(s)$ with positive real part, this assumption is equivalent to assume that $P_{\bar{G}} = P_G$;
- there are no cancellations of poles and zeros of $R(s)$ and $G(s)$ with non negative real part.

Then, a sufficient condition guaranteeing that the feedback system is asymptotically stable when the true transfer function of the process is $G(s)$ is given by

$$|R(j\omega)\Delta G_a(j\omega)| < |1 + L(j\omega)| = |S(j\omega)|^{-1} \ ,\forall\omega$$

where $L(s) = R(s)\bar{G}(s)$ is the nominal loop transfer function and $S(s)$ is computed on $L(s)$.

This corresponds to

$$|S(j\omega)| < \frac{1}{|R(j\omega)\Delta G_a(j\omega)|} \ ,\forall\omega$$

In fact, this condition, together with the adopted assumptions, guarantees that the number of encirclements of the critical point -1 of the Nyquist diagram of the nominal sensitivity function is equal to the one of the perturbed sensitivity function, see Figure 3.9.

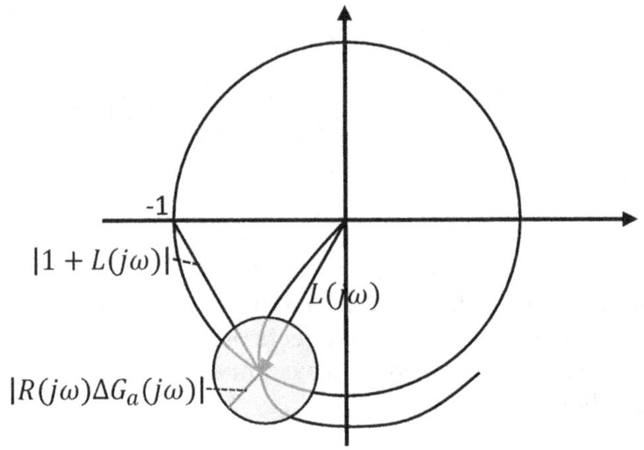

Figure 3.9: Robustness conditions for additive uncertainty.

Therefore, also in this case, it is required that $|S(j\omega)|$ is "small", at least at those frequencies where $|\Delta G_a(j\omega)|$ is "big". This motivates the need to require a "small" M_S.

A different approach, leading to the same conclusions, can be followed based on the small gain theorem. To this end, consider the perturbed closed-loop model represented in Figure 3.10 and note that $\phi = -R(s)(\eta + \bar{G}(s)\phi)$, that is

$$\phi = -\frac{R(s)}{1+\bar{G}(s)R(s)}\eta$$

Therefore, an equivalent scheme can be drawn as in Figure 3.11. Since $\frac{R(s)}{1+\bar{G}(s)R(s)}$ and $\Delta G_a(s)$ are asymptotically stable, the small gain theorem can be applied to conclude that a sufficient condition for closed-loop stability is

$$\|RS\Delta G_a\|_\infty < 1$$

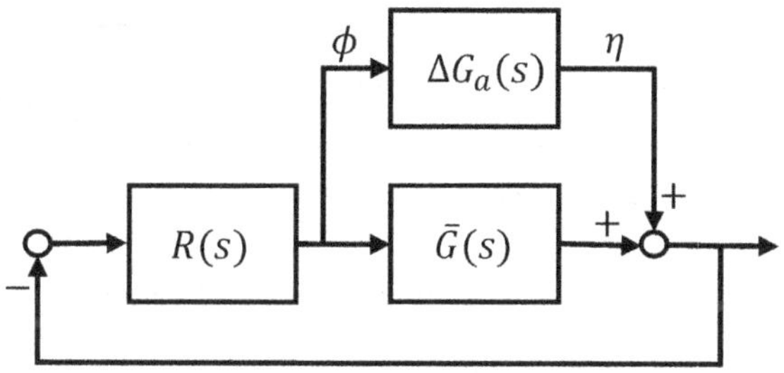

Figure 3.10: Additive uncertainty: Closed-loop system.

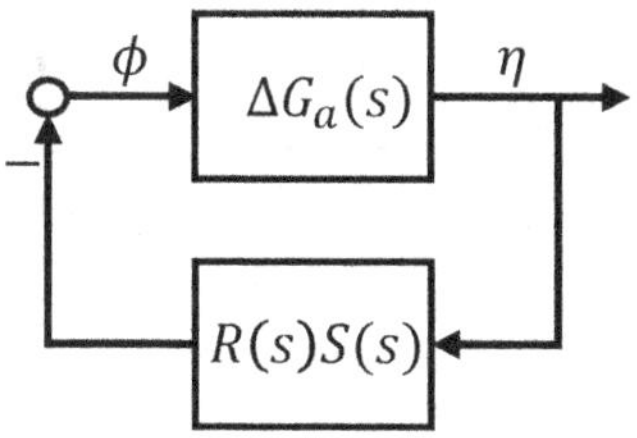

Figure 3.11: Additive uncertainty: closed-loop system redesign.

3.5 Robustness with respect to multiplicative uncertainty

Let the system be described by

$$G(s) = \bar{G}(s)(1 + \Delta G_m(s))$$

where $\bar{G}(s)$ is the nominal transfer function. Moreover, assume that:

- the closed-loop nominal system is asymptotically stable;
- $\Delta G_m(s)$ is an asymptotically stable transfer function. Denoting by $P_{\bar{G}}$ and P_G the number of poles of $\bar{G}(s)$ and $G(s)$ with positive real part, this assumption is equivalent to assume that $P_{\bar{G}} = P_G$.

In order to find conditions guaranteeing the asymptotic stability of the perturbed closed-loop system, consider the block manipulations shown in Figure 3.12-3.13 where $T(s)$ is the complementary sensitivity in nominal conditions.

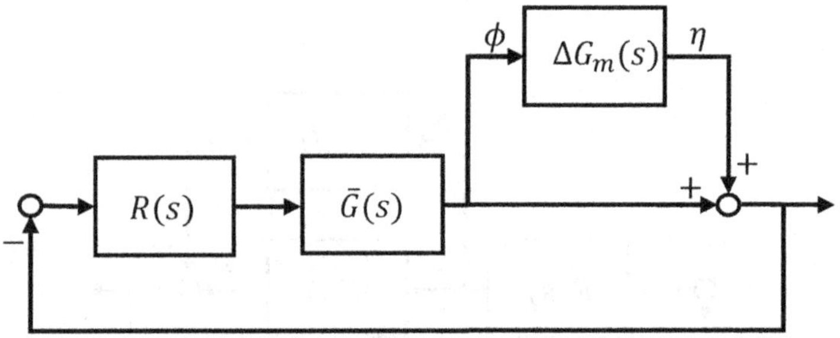

Figure 3.12: Multiplicative uncertainty: close-loop system.

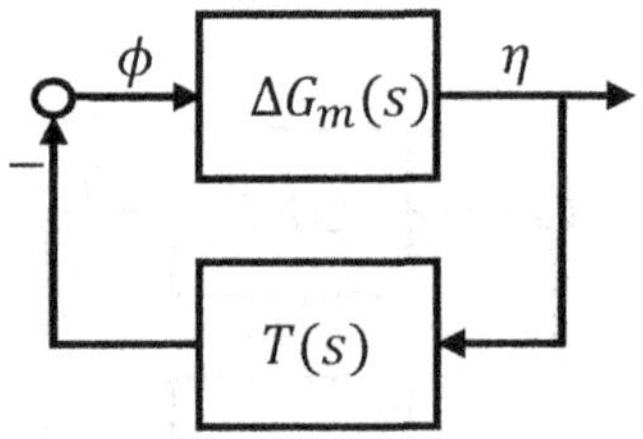

Figure 3.13: Multiplicative uncertainty: close-loop system redesign.

By assumption, both $T(s)$ and $\Delta G_m(s)$ are asymptotically stable transfer functions, as such, also the loop transfer function $T(s)\Delta G_m(s)$ is asymptotically stable. In view of the Nyquist theorem, the closed-loop system is asymptotically stable if and only if the number of encirclements of -1 of $T(j\omega)\Delta G_m(j\omega)$ is equal to zero. This is guaranteed if

$$|T(j\omega)\Delta G_m(j\omega)| < 1, \;\; \forall\omega$$

or

$$\|T\Delta G_m\|_\infty < 1$$

From these results, it is possible to conclude that limiting the value of M_S or M_T guarantees robustness for additive or multiplicative uncertainties.

3.6 Design specifications on the sensitivity and complementary sensitivity functions

Now define by ω_B the frequency where $|S(j\omega)|$ crosses $1/\sqrt{2}$ (-3dB) from below and by ω_{BT} the frequency where $|T(j\omega)|$ crosses $1/\sqrt{2}$ (-3dB) from above (see Figure 3.14).

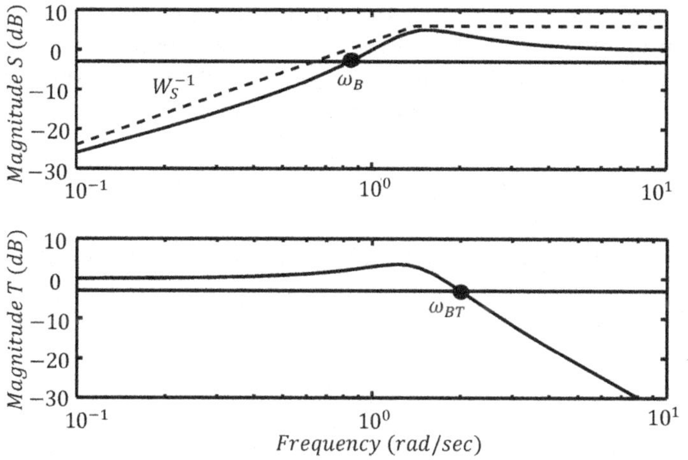

Figure 3.14: Bode diagram of the sensitivity and complementary sensitivity functions.

It is possible to prove that, if $\varphi_m < 90^o$, one has

$$\omega_B < \omega_c < \omega_{BT}$$

Therefore, suitable design specifications on $S(s)$ can be:

- minimum frequency ω_B;
- small or null asymptotic error for constant reference signals ($|S(j\omega)|$ small or Bode diagram of $|S(j\omega)|$ with shape $+1$ at low frequency);
- shape of $|S(j\omega|$;
- $M_S \leq \bar{M}_S$.

These specifications can be formally stated by requiring that

$$|S(j\omega)| < \frac{1}{|W_S(j\omega)|}, \quad \forall\omega$$

where $W_S(j\omega)$ is a "shaping function" defined by the designer, which can be seen as the inverse of the required (ideal) sensitivity function, see Figure 3.14. This requirement can be reformulated as

$$\|W_S S\|_\infty < 1$$

A possible choice of $W_S(s)$ is

$$W_S(s) = \frac{s/M + \omega_B}{s + A\omega_B},$$

where $A \ll 1$ is the desired attenuation in the band of interest and M is the required bound on the H_∞ norm of the sensitivity function.

Similar arguments could be used to specify performance and robust stability requirements in terms $T(s)$. Specifically,

$$|T(j\omega)| < \frac{1}{|W_T(j\omega)|}, \quad \forall\omega \Leftrightarrow \|W_T T\|_\infty < 1$$

where $W_T(s)$ is a transfer function suitably selected by the designer.

A possible choice is

$$W_T(s) = \frac{s + \omega_{BT}/M}{As + \omega_{BT}},$$

where A and M can be selected according to the previous considerations.

Finally, recall that the control variable is given by

$$U(s) = \frac{R(s)}{1 + R(s)G(s)} Y^o(s) = K(s)Y^o(s)$$

Hence, in order to limit the control action, one can set

$$|K(j\omega)| < \frac{1}{|W_K(j\omega)|}, \quad \forall\omega \Leftrightarrow \|W_K K\|_\infty < 1$$

where, again, $W_K(s)$ is a suitable design parameter, i.e. a transfer function to be properly selected by the designer.

The shaping function $W_S(s)$, $W_T(s)$, $W_K(s)$ must be selected as asymptotically stable systems. This contrasts with the need to use a function $W_S(s)$ with infinite gain at low frequency, so as to obtain a loop transfer function $L(s)$, and a regulator $R(s)$, with integral action. This problem can be solved by using a function $W_S(s)$ with very high (but finite) gain at low frequency. The resulting regulator will have a pole near the origin, which could be approximated by an integrator in the final design phase.

Given the shaping functions $W_S(s)$, $W_T(s)$, and $W_K(s)$, the regulator can be obtained (at least conceptually) by imposing

$$\|W_S S\|_\infty < 1, \quad \|W_T T\|_\infty < 1, \quad \|W_K K\|_\infty < 1$$

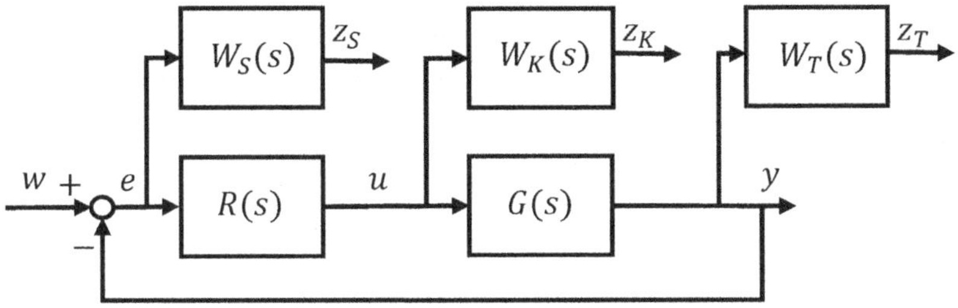

Figure 3.15: System with shaping functions.

Alternatively, letting

$$G_{zw}(s) = \begin{bmatrix} W_S(s)S(s) \\ W_T(s)T(s) \\ W_K(s)K(s) \end{bmatrix}$$

the control synthesis can be completed by minimizing

$$\|G_{zw}\|_\infty = \sup_\omega \bar{\sigma}(G_{zw}(j\omega))$$

If the resulting regulator is such that $\|G_{zw}\|_\infty < \gamma$ one has

$$\|W_S S\|_\infty < \gamma, \quad \|W_T T\|_\infty < \gamma, \quad \|W_K K\|_\infty < \gamma$$

This synthesis method is usually called H_∞ *control.*

A different approach, usually called H_2 *control*, consists of minimizing

$$\begin{aligned} J &= \frac{1}{2\pi}\int_{-\infty}^{+\infty} \left(|W_S(j\omega)S(j\omega)|^2 + |W_T(j\omega)T(j\omega)|^2 + |W_K(j\omega)K(j\omega)|^2\right) d\omega \\ &= \frac{1}{2\pi}\int_{-\infty}^{+\infty} |G_{zw}(j\omega)|^2 \, d\omega \end{aligned}$$

With reference to Figure 3.15, note that, by letting

$$z = \begin{bmatrix} z_S \\ z_K \\ z_T \end{bmatrix}, \quad w = y^o$$

$G_{zw}(s)$ is the transfer function from w to z when the enlarged system made by $G(s)$ and by the functions $W_S(s)$, $W_T(s)$, $W_K(s)$ is fed-back with the regulator $R(s)$.

In general, define by w the vector of the exogenous signals, for example

$$w = \begin{bmatrix} d \\ y^o \\ n \end{bmatrix}$$

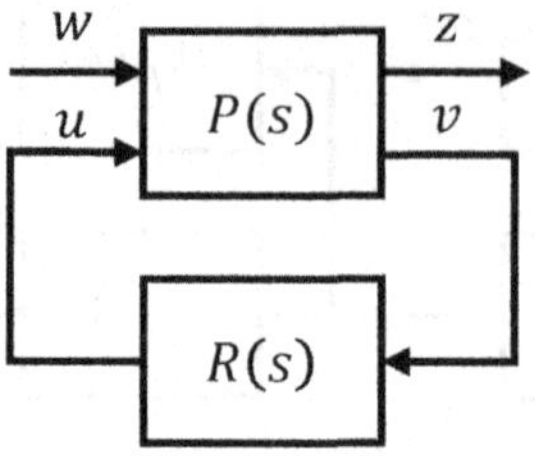

Figure 3.16: Scheme for H_2 and H_{inf} control.

by u the control variable, by z the set of *performance variables* (for instance y and u), and by v the measured variable. Then, the H_∞, or H_2, control problem consists of minimizing, with respect to $R(s)$, the norm of the transfer function from w to z in the enlarged system depicted in Figure 3.16.

4

Multivariable poles and zeros

In this chapter, the definitions of poles and zeros are extended to multivariable continuous time linear systems. Most of the results are valid also for discrete time systems.

4.1 Preliminaries

Consider the following linear time invariant system

$$\begin{aligned} \dot{x}(t) &= Ax(t) + Bu(t) \\ y(t) &= Cx(t) + Du(t) \end{aligned}$$

with $x \in R^n, u \in R^m, y \in R^p$.

In the Laplace domain, the system is described by the following input-output relation

$$Y(s) = G(s)U(s)$$

where

$$G(s) = C(sI - A)^{-1}B + D = \begin{bmatrix} G_{11}(s) & \dots & G_{1m}(s) \\ \vdots & \ddots & \vdots \\ G_{p1}(s) & \dots & G_{pm}(s) \end{bmatrix}$$

is a matrix of transfer functions. The transformation from the state-space model to the transfer function is unique and easily computable through the previous formula. On the contrary, the realization problem, that is the problem of finding a state space representation of $G(s)$, is trivial only for *SISO* (Single Input Single Output) systems, where for example the controllable, or reachable, and observable canonical forms are available. For *MIMO* (Multi Input Multi Output) systems, there exist conceptually similar canonical forms, but they are much more difficult to derive. It is important to recall that the transfer function matrix $G(s)$ describes only the reachable and observable part of the system. Hence, when in the following the system will be described in the Laplace domain, it will be implicitly assumed that it is in minimal form, i.e. it is completely reachable and observable.

4.2 Poles

Definition 4.2.1. For systems in minimal form, the poles, including their multiplicity, coincide with the eigenvalues of the matrix A, i.e. with the roots of the characteristic equation

$$\phi(s) = \det(sI - A) = 0$$

4.2.1 Computation of the poles from $G(s)$

The poles can be computed directly from the knowledge of the transfer function. In fact, the following result holds.

Theorem 4.2.1. *The characteristic polynomial $\phi(s)$ associated with a minimal realization of a system with transfer function matrix $G(s)$ is the least common denominator of all the non null minors of any order of $G(s)$.*

Example 4.2.1. Consider the system described by

$$G(s) = \begin{bmatrix} \frac{1}{s+1} & 0 \\ 0 & \frac{3}{s+1} \end{bmatrix}$$

the first order minors are

$$\frac{1}{s+1}; \frac{3}{s+1}$$

the second order minor is

$$\frac{1}{s+1}\frac{3}{s+1} = \frac{3}{(s+1)^2}$$

Then, the system has two poles in $s = -1$.

Example 4.2.2. Consider the system

$$G(s) = \begin{bmatrix} \frac{1}{s+1} & -\frac{1}{s+1} \\ \frac{1}{s+1} & \frac{s}{s+1} \end{bmatrix}$$

the first order minors are

$$\frac{1}{s+1}; -\frac{1}{s+1}; \frac{1}{s+1}; \frac{s}{s+1}$$

the second order minor is

$$\frac{1}{s+1}\frac{s}{s+1} + \frac{1}{s+1}\frac{1}{s+1} = \frac{s+1}{(s+1)^2} = \frac{1}{s+1}$$

Then, the system is first order and has only one pole in $s = -1$.

A possible minimal state-space realization is given by the following first order system

$$\begin{aligned}\dot{x}(t) &= -x(t) + \begin{bmatrix} 1 & -1 \end{bmatrix} \begin{bmatrix} u_1(t) \\ u_2(t) \end{bmatrix} \\ \begin{bmatrix} y_1(t) \\ y_2(t) \end{bmatrix} &= \begin{bmatrix} 1 \\ 1 \end{bmatrix} x(t) + \begin{bmatrix} 0 & 0 \\ 0 & 1 \end{bmatrix} \begin{bmatrix} u_1(t) \\ u_2(t) \end{bmatrix}\end{aligned}$$

which shows that only one state is required to fully describe the original system specified by $G(s)$.

Example 4.2.3. Consider the system

$$G_3(s) = \begin{bmatrix} \frac{1}{(s+1)} & 0 & \frac{(s-1)}{(s+1)(s+2)} \\ \frac{-1}{(s-1)} & \frac{1}{(s+2)} & \frac{1}{(s+2)} \end{bmatrix}$$

the first order minors are

$$\frac{1}{(s+1)}; \frac{(s-1)}{(s+1)(s+2)}; \frac{-1}{(s-1)}; \frac{1}{(s+2)}; \frac{1}{(s+2)}$$

while the second order minors are:
A) by removing column 2

$$\begin{aligned}M_2 &= \det \begin{bmatrix} \frac{1}{s+1} & \frac{s-1}{(s+1)(s+2)} \\ -\frac{1}{s-1} & \frac{1}{(s+2)} \end{bmatrix} \\ &= \frac{1}{s+1}\frac{1}{s+2} + \frac{1}{(s+1)(s+2)} \\ &= \frac{2}{(s+1)(s+2)}\end{aligned}$$

B) by removing column 3

$$M_3 = \det \begin{bmatrix} \frac{1}{s+1} & 0 \\ -\frac{1}{s-1} & \frac{1}{(s+2)} \end{bmatrix} = \frac{1}{(s+1)(s+2)}$$

C) by removing column 1

$$\begin{aligned}M_1 &= \det \begin{bmatrix} 0 & \frac{s-1}{(s+1)(s+2)} \\ \frac{1}{s+2} & \frac{1}{(s+2)} \end{bmatrix} \\ &= -\frac{(s-1)}{(s+1)(s+2)^2}\end{aligned}$$

Then, the minimum common denominator is

$$\phi(s) = (s+1)(s+2)^2(s-1)$$

and the system has 4 poles in -1, 1, -2 -2.

Remark. From the previous examples it is possible to note that poles are also poles of the individual transfer functions corresponding to the elements of $G(s)$, but their multiplicity cannot be easily derived from the analysis of the elements of $G(s)$.

4.3 Zeros of *SISO* systems

In order to better understand the meaning of the zeros of *MIMO* systems, in the following the blocking property of the zeros of *SISO* systems is introduced.

4.3.1 Blocking property of the zeros

First consider a *SISO* system in minimal form described by

$$\begin{array}{l}\dot{x}(t) = Ax(t) + Bu(t) \\ y(t) = Cx(t) + Du(t)\end{array}, \quad G(s) = C(sI - A)^{-1}B + D$$

with input $u(t) = u_0 e^{\lambda t}$, whose Laplace transform is $U(s) = \frac{u_0}{s-\lambda}$. The following result holds.

Theorem 4.3.1. *If λ is not an eigenvalue of A, there exists an initial state x_0 such that, given the input $u_0 e^{\lambda t}$, the output has the same exponential behavior, i.e. $y(t) = y_0 e^{\lambda t}$.*

Proof. Let assume that there exists an initial state x_0 such that the state evolution is $x(t) = x_0 e^{\lambda t}$. From the system equation it follows that

$$\dot{x}(t) = \lambda x_0 e^{\lambda t} = Ax_0 e^{\lambda t} + Bu_0 e^{\lambda t}$$

that is

$$(\lambda I - A)\, x_0 = Bu_0$$

Then x_0 exists and is unique, since by assumption λ is not an eigenvalue of A, i.e.

$$x_0 = (\lambda I - A)^{-1} Bu_0$$

and

$$x(t) = (\lambda I - A)^{-1} Bu_0 e^{\lambda t}$$

Therefore

$$\begin{aligned} y(t) &= Cx(t) + Du(t) \\ &= \left[C\,(\lambda I - A)^{-1} B + D\right] u_0 e^{\lambda t} \\ &= G(\lambda) u_0 e^{\lambda t} \end{aligned}$$

Note that if λ is a zero of $G(s)$, $G(\lambda) = 0$. Hence, in this case there exists an initial state x_0 such that, given the exponential input $u_0 e^{\lambda t}$, the output is null at any t. The input is then blocked by the system and the following result holds.

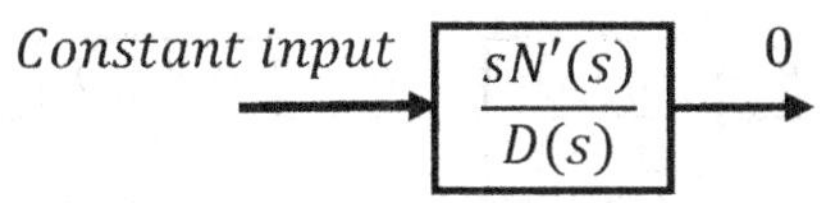

Figure 4.1: System with derivative action (open loop).

Theorem 4.3.2. *If λ is a zero of $G(s)$, given u_0 there exits an initial state x_0 such that the output associated with the initial state x_0 and the input $u(t) = u_0 e^{\lambda t}$ is null for any $t \geq 0$.*

Assuming that the system is asymptotically stable, the zero blocking property can be trivially proven, at least asymptotically. In fact, consider the input $U(s) = \frac{u_0}{s-\lambda}$ and suppose that $G(s)$ is asymptotically stable and given by

$$G(s) = \frac{(s-\lambda)N'(s)}{D(s)}$$

then

$$Y(s) = G(s)U(s) = \frac{(s-\lambda)N'(s)}{D(s)} \frac{u_0}{s-\lambda} = \frac{N'(s)}{D(s)} u_0$$

In view of the asymptotic stability of $G(s)$, it is possible to use the final value theorem to conclude that

$$\lim_{t \to \infty} y(t) = \lim_{s \to 0} sY(s) = \lim_{s \to 0} s \frac{N'(s)}{D(s)} u_0 = 0$$

Hence, an asymptotically null output is obtained also with an exponential divergent input ($\lambda > 0$).

4.3.2 Role of the zeros in the design of the control system

The analysis of the zeros is important to find possible limits of the performance of the control system.

Example 4.3.1. Consider the asymptotically stable system, represented in Figure 4.1, described by $G(s) = s\frac{N'(s)}{D(s)}$, that is characterized by a derivative action. From the above results, it is possible to conclude that, with a constant input $u(t) = u_0$, $t \geq 0$, $U(s) = \frac{u_0}{s}$ (in analogy with the previous considered case with $\lambda = 0$), the output asymptotically tends to zero. This result is obvious because, as shown in Figure 4.1, the derivative action on the constant input gives an asymptotically null output.

For closed-loop systems, see Figure 4.2, similar conclusions can be drawn; in fact with $R(s) = \frac{N_R(s)}{D_R(s)}$ it follows that

$$Y(s) = \frac{R(s)G(s)}{1 + R(s)G(s)} Y^0(s) = \frac{sN'(s)N_R(s)}{D_R(s)D(s) + N_R(s)sN'(s)} Y^0(s)$$

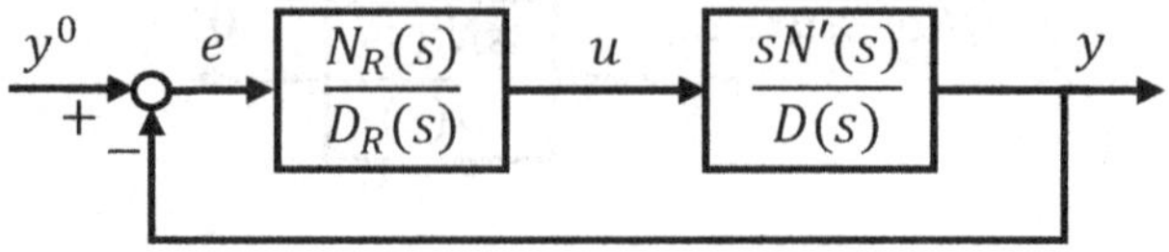

Figure 4.2: System with derivative action (closed-loop).

The non asymptotically stable open-loop zeros, roots of $sN'(s)$, are also zeros of the closed-loop system (recall that cancellations of these singularities are not allowed, in particular for $s = 0$). This means that there is a derivative action also in the transfer function from the set-point to the output. Hence, the output cannot asymptotically reach any value different from zero. This shows that the null zeros, with their blocking property, introduce a limitation on the static performance.

Considering dynamic performance limitations, recall that, in SISO systems, the zeros with positive real part give a negative phase contribution (see Section 3.1.1) and hence limit the achievable bandwidth of the closed-loop system and the speed of response.

4.4 Zeros of *MIMO* systems

A more general definition of zeros, valid also for *MIMO* systems, is now introduced.

4.4.1 Invariant zeros

Note that, for $x_0 = 0$, the system in the Laplace domain can be written as

$$\begin{aligned} sX(s) &= AX(s) + BU(s) \\ Y(s) &= CX(s) + DU(s) \end{aligned}$$

or

$$\begin{bmatrix} sI - A & -B \\ C & D \end{bmatrix} \begin{bmatrix} X(s) \\ U(s) \end{bmatrix} = \begin{bmatrix} 0 \\ Y(s) \end{bmatrix}$$

Define the *system matrix* $P(s)$ as

$$P(s) = \begin{bmatrix} sI - A & -B \\ C & D \end{bmatrix}$$

Definition 4.4.1. Given a matrix $P(s)$, its *normal rank* is its rank for any value of s, except for at most a finite number of singularities.

Definition 4.4.2. The invariant zeros of a system are the values of s such that the rank of the system matrix $P(s)$ is lower than the normal rank.

Example 4.4.1. To understand the meaning of invariant zeros, first consider a *SISO* system given by

$$G(s) = \frac{\beta_3 s^3 + \beta_2 s^2 + \beta_1 s + \beta_0}{s^4 + \alpha_3 s^3 + \alpha_2 s^2 + \alpha_1 s + \alpha_0}$$

The system can be described by a state-space model in the following controllable canonical form

$$\begin{aligned} A &= \begin{bmatrix} 0 & 1 & 0 & 0 \\ 0 & 0 & 1 & 0 \\ 0 & 0 & 0 & 1 \\ -\alpha_0 & -\alpha_1 & -\alpha_2 & -\alpha_3 \end{bmatrix}, \quad B = \begin{bmatrix} 0 \\ 0 \\ 0 \\ 1 \end{bmatrix} \\ C &= \begin{bmatrix} \beta_0 & \beta_1 & \beta_2 & \beta_3 \end{bmatrix}, \quad D = 0 \end{aligned}$$

whose system matrix is

$$P(s) = \begin{bmatrix} s & -1 & 0 & 0 & 0 \\ 0 & s & -1 & 0 & 0 \\ 0 & 0 & s & -1 & 0 \\ \alpha_0 & \alpha_1 & \alpha_2 & s+\alpha_3 & -1 \\ \beta_0 & \beta_1 & \beta_2 & \beta_3 & 0 \end{bmatrix}$$

To compute the determinant of $P(s)$, note that

$$\det P(s) = -1 * (-1)^{5+4} \det P'(s) = \det P'(s)$$

where

$$\begin{aligned} \det P'(s) &= \det \begin{bmatrix} s & -1 & 0 & 0 \\ 0 & s & -1 & 0 \\ 0 & 0 & s & -1 \\ \beta_0 & \beta_1 & \beta_2 & \beta_3 \end{bmatrix} \\ &= -1 * (-1)^{4+3} \det \begin{bmatrix} s & -1 & 0 \\ 0 & s & -1 \\ \beta_0 & \beta_1 & \beta_2 \end{bmatrix} \\ &\quad + \beta_3 * (-1)^{4+4} \det \begin{bmatrix} s & -1 & 0 \\ 0 & s & -1 \\ 0 & 0 & s \end{bmatrix} \\ &= -1 * (-1)^{3+2} \det \begin{bmatrix} s & -1 \\ \beta_0 & \beta_1 \end{bmatrix} + \beta_2 (-1)^{3+3} \det \begin{bmatrix} s & -1 \\ 0 & s \end{bmatrix} + \beta_3 s^3 \\ &= s\beta_1 + \beta_0 + \beta_2 s^2 + \beta_3 s^3 \end{aligned}$$

Finally

$$\det(P(s)) = \beta_3 s^3 + \beta_2 s^2 + s\beta_1 + \beta_0$$

Hence, in this case, and in general for *SISO* systems, the invariant zeros coincide with the zeros of the transfer function $G(s)$.

Coming back to *MIMO* systems, it is clear that the *invariant zeros are not the zeros of the single elements of the transfer function matrix* $G(s)$, but they enjoy the blocking property, as stated in the following theorem.

Theorem 4.4.1. *If λ is an invariant zero, there exist an initial state x_0 and a vector u_0 such that, given an input $u(t) = u_0 e^{\lambda t}$, the output is $y(t) = 0$, $t \geq 0$.*

Proof. For simplicity, we assume that λ is an invariant zero not coinciding with a pole of $G(s)$. Then, there exists $\begin{bmatrix} x_0 \\ u_0 \end{bmatrix} \neq 0$ such that

$$\begin{bmatrix} \lambda I - A & -B \\ C & D \end{bmatrix} \begin{bmatrix} x_0 \\ u_0 \end{bmatrix} = 0$$

that is

$$\begin{aligned} (\lambda I - A)\, x_0 &= B u_0 \\ C x_0 + D u_0 &= 0 \end{aligned}$$

Given $U(s) = \frac{u_0}{s-\lambda}$, it is also true that

$$\begin{aligned} X(s) &= (sI - A)^{-1} \left[x_0 + BU(s)\right] \\ &= (sI - A)^{-1} \left[x_0 + \frac{Bu_0}{s - \lambda}\right] \end{aligned}$$

But, since λ is an invariant zero,

$$(\lambda I - A)\, x_0 = B u_0$$

that is

$$(\lambda I - A) \frac{1}{s - \lambda} x_0 = \frac{B}{s - \lambda} u_0$$

hence

$$\begin{aligned} X(s) &= (sI - A)^{-1} \left[x_0 + (\lambda I - A) \frac{1}{s - \lambda} x_0\right] \\ &= (sI - A)^{-1} \left[\frac{sIx_0 - \lambda I x_0 + \lambda I x_0 - A x_0}{s - \lambda}\right] \\ &= (sI - A)^{-1} \left[\frac{(sI - A)\, x_0}{s - \lambda}\right] = \frac{x_0}{s - \lambda} \end{aligned}$$

Then, it follows that

$$\begin{aligned} Y(s) &= CX(s) + DU(s) \\ &= C\frac{x_0}{s-\lambda} + D\frac{u_0}{s-\lambda} \end{aligned}$$

or

$$(s-\lambda)\,Y(s) = Cx_0 + Du_0 = 0$$

In the time domain one has

$$\dot{y}(t) = \lambda y(t)$$

whose solution is

$$y(t) = e^{\lambda t} y(0)$$

but, since $y(0) = 0$, one has $y(t) = 0$, $t \geq 0$.

Remark. Comparing the results of Theorems 4.3.2 and 4.4.1, it is apparent that, while in the *SISO* case for any input $u_0 e^{\lambda t}$ it is possible to find the initial state x_0 such that $y(t) = 0$, in the ***MIMO*** case the pair (x_0, u_0) must be jointly selected in order to obtain a zero output at any time.

4.4.2 Computation of the invariant zeros from $G(s)$

As well as the poles, also the invariant zeros can be computed directly from the transfer function matrix $G(s)$.

Definition 4.4.3. The polynomial $z(s)$ of the invariant zeros of $G(s)$ is the polynomial with roots coinciding with all and only the invariant zeros of $G(s)$.

Theorem 4.4.2. *The polynomial $z(s)$ of the invariant zeros of $G(s)$ is the greatest common divisor of all the numerators of all the minors of order r of $G(s)$, where r is the normal rank of $G(s)$, assuming that these minors are written so that they have the polynomial $\phi(s)$ of the poles at the denominator.*

Example 4.4.2. The characteristic polynomial of the system described by

$$G(s) = \frac{1}{(0.2s+1)(1+s)} \begin{bmatrix} 1 & 1 \\ 1+2s & 2 \end{bmatrix}$$

is

$$\phi(s) = (0.2s+1)^2 (1+s)^2$$

and its normal rank is 2, since

$$detG(s) = \frac{2-1-2s}{\phi(s)} = \frac{1-2s}{\phi(s)}$$

Then, the unique zero is equal to 0.5 (real and positive), note that the individual transfer functions do not have zeros with positive real part, but the invariant zero is positive.

Example 4.4.3. The system

$$G(s) = \begin{bmatrix} \frac{s+1}{(s+2)(s+3)} & \frac{4}{(s+2)(s+3)} \\ \frac{0.5}{(s+2)(s+3)} & \frac{2}{(s+2)(s+3)} \end{bmatrix}$$

has normal rank 2, characteristic polynomial

$$\phi(s) = (s+2)^2 (s+3)^2$$

and

$$detG(s) = \frac{2s+2-2}{\phi(s)} = \frac{2s}{\phi(s)}$$

Hence there is an invariant zero in $s = 0$, that is a derivative action. Note that at the steady state it is possible to impose a constant value to anyone of the outputs (the individual transfer functions of $G(s)$ do not have zeros at the origin), but it is not possible to contemporary impose arbitrary values to both the outputs. In fact, the static gain is

$$G(0) = \begin{bmatrix} 1 & 4 \\ 0.5 & 2 \end{bmatrix} \frac{1}{6}$$

and $det(G(0)) = 0$. At the steady state, in order to have constat arbitrary values $\bar{y}_1$ and $\bar{y}_2$ of the output, it would be necessary to find constant inputs $\bar{u}_1$ and $\bar{u}_2$ such that

$$\begin{bmatrix} \bar{y}_1 \\ \bar{y}_2 \end{bmatrix} = G(0) \begin{bmatrix} \bar{u}_1 \\ \bar{u}_2 \end{bmatrix}$$

that is

$$\begin{bmatrix} \bar{u}_1 \\ \bar{u}_2 \end{bmatrix} = G(0)^{-1} \begin{bmatrix} \bar{y}_1 \\ \bar{y}_2 \end{bmatrix}$$

clearly not possible because $G(0)$ is not invertible.

Example 4.4.4. The system

$$G(s) = \begin{bmatrix} \frac{1}{(s+1)} & 0 & \frac{(s-1)}{(s+1)(s+2)} \\ \frac{-1}{(s-1)} & \frac{1}{(s+2)} & \frac{1}{(s+2)} \end{bmatrix}$$

has characteristic polynomial

$$\phi(s) = (s+1)(s+2)^2(s-1)$$

The second order minors, written so that the denominator is equal to the characteristic polynomial, are

$$M_2 = \frac{2}{(s+1)(s+2)} = \frac{2(s+2)(s-1)}{(s+1)(s+2)^2(s-1)}$$

$$M_1 = \frac{-(s-1)}{(s+1)(s+2)^2} = \frac{-(s-1)^2}{(s+1)(s+2)^2(s-1)}$$

$$M_3 = \frac{1}{(s+1)(s+2)} = \frac{(s+2)(s-1)}{(s+1)(s+2)^2(s-1)}$$

Then, the greatest common divisor is

$$z(s) = (s-1)$$

and the system has one zero in 1.

Example 4.4.5. The system

$$G(s) = \frac{1}{(s+2)} \begin{bmatrix} (s-1) & 4 \\ 4.5 & 2(s-1) \end{bmatrix}$$

has characteristic polynomial

$$\phi(s) = s+2$$

and the unique minor of second order is

$$\det(G(s)) = \frac{2(s-1)^2 - 18}{(s+2)^2} = \frac{2(s^2-2s-8)}{(s+2)^2} = \frac{2(s-4)(s+2)}{(s+2)^2} = \frac{2(s-4)}{(s+2)}$$

Hence, the system has a pole in -2 and a zero in 4 (note that 4 is not a zero of any individual transfer function).

5 Multivariable systems analysis

In the following, some methods for the analysis of single-input single-output dynamical systems are extended to the multivariable (*MIMO*) case. First, *open-loop* systems will be considered, then *closed-loop* systems will be analyzed.

5.1 Open loop systems

Since the stability of a linear, time invariant dynamic system depends only on the dynamic matrix A, also in the *MIMO* case the system is asymptotically stable if and only if the eigenvalues of A have negative real part.

5.1.1 Frequency response

First recall that, given an asymptotically stable *SISO* system with transfer function $G(s)$ and input signal $u(t) = Usin(\bar{\omega}t)$, its output asymptotically tends to $y(t) = Ysin(\bar{\omega}t + \alpha)$, where $Y = U\,|G(j\bar{\omega})|$ and $\alpha = \arg(G(j\bar{\omega}))$. Therefore, in general

$$|G(j\bar{\omega})| = \frac{|Y(j\bar{\omega})|}{|U(j\bar{\omega})|}$$

can be interpreted as the "gain" of the system at frequency $\bar{\omega}$ (see also Section 2.1 for different definitions of the system gain).

For *MIMO* systems, this result can be used by considering individual input-output pairs. However, in order to extend the previous concepts, recall that in Chapter 2 the gain of the system at a given frequency ω has been defined as

$$\frac{\|Y(j\omega)\|_2}{\|U(j\omega)\|_2} = \frac{\|G(j\omega)U(j\omega)\|_2}{\|U(j\omega)\|_2}$$

Note that the gain does not depend only on ω, but also on the input U "direction". To estimate the value of the gain at different values of ω, recall also the following relation

$$\underline{\sigma}(G(j\omega)) \leq \frac{\|G(j\omega)U(j\omega)\|_2}{\|U(j\omega)\|_2} \leq \bar{\sigma}(G(j\omega))$$

Hence, the system gain, at any ω, lies between the minimum singular value $\underline{\sigma}$ and the maximum singular value $\bar{\sigma}$ at that frequency. Then, as in the *SISO* case the Bode diagrams of $G(j\omega)$ are a fundamental analysis tool, in the *MIMO* case it is possible to draw the diagrams of $\underline{\sigma}(G(j\omega))$ and $\bar{\sigma}(G(j\omega))$ as functions of ω. To obtain a "small" output y at a given $\bar{\omega}$ regardless of the input direction, that is to guarantee attenuation at $\bar{\omega}$, one must require that $\bar{\sigma}(G(j\bar{\omega}))$ is small. Conversely, if it is required that the system amplifies at $\bar{\omega}$, $\underline{\sigma}(G(j\bar{\omega}))$ must be "large".

Remark. According to the previous considerations, the singular values $\underline{\sigma}(G(j\omega))$ and $\bar{\sigma}(G(j\omega))$ of the system are also called "principal gains".

Remark. The condition number is defined as

$$\gamma(G(j\omega)) = \frac{\bar{\sigma}(G(j\omega))}{\underline{\sigma}(G(j\omega))}$$

It is customary to assume that a system with a condition number close to 1 is "easy" to control, since it is not very sensitive to the direction of the applied input.

Example 5.1.1. Consider a static system with two inputs, two outputs, and described by the matrix

$$G = \begin{bmatrix} 1 & 2 \\ 3 & 4 \end{bmatrix}$$

and the following inputs with norm equal to one

$$U_1 = \begin{bmatrix} 1 \\ 0 \end{bmatrix}, U_2 = \begin{bmatrix} 0 \\ 1 \end{bmatrix}, U_3 = \begin{bmatrix} 0.707 \\ 0.707 \end{bmatrix}, U_4 = \begin{bmatrix} 0.707 \\ -0.707 \end{bmatrix}, U_5 = \begin{bmatrix} 0.8 \\ -0.6 \end{bmatrix}$$

The corresponding outputs are

$$Y_1 = \begin{bmatrix} 1 \\ 3 \end{bmatrix}, Y_2 = \begin{bmatrix} 2 \\ 4 \end{bmatrix}, Y_3 = \begin{bmatrix} 2.12 \\ 4.95 \end{bmatrix}, Y_4 = \begin{bmatrix} -0.707 \\ -0.707 \end{bmatrix}, Y_5 = \begin{bmatrix} -0.4 \\ 0 \end{bmatrix}$$

with norms

$$\|Y_1\|_2 = 3.16, \|Y_2\|_2 = 4.47, \|Y_3\|_2 = 5.38, \|Y_4\|_2 = 1.00, \|Y_5\|_2 = 0.4$$

It is apparent that, since the inputs have the same norm, their direction strongly influences the norm of the output.

To verify this result, by means of the formula reported in Appendix A for the computation of the singular values of a two-input, two-output system, one obtains

$$\bar{\sigma}(G) = 5.47; \quad \underline{\sigma}(G) = 0.37$$

As expected, recalling that the considered inputs U_i, $i = 1, \ldots, 5$, have norm 1, the derived gains ($\|Y_i\|_2$) are between the minimum singular value $\underline{\sigma}(G)$ and the maximum one $\bar{\sigma}(G)$.

5.2 Closed-loop systems

5.2.1 Block diagrams manipulation

First note that, given two matrices G and K of proper dimensions, in general one has

$$GK \neq KG$$

even when the two products make sense. Therefore, when manipulating block diagrams, it is necessary to pay attention to the order of the terms in the products.

Series or cascade systems

Given the system reported in Figure 5.1, where $G_1(s)$ and $G_2(s)$ are transfer functions of proper dimensions, it results

$$Y_2(s) = G_2(s)U_2(s) = G_2(s)G_1(s)U_1(s)$$

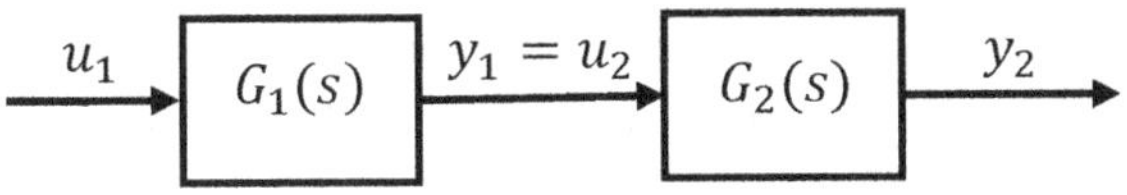

Figure 5.1: Systems in series configuration.

Feedback schemes

Given the closed-loop system reported in Figure 5.2, where $y_1 \in R^p$ and $y_2 \in R^m$, it is possible to write

$$\begin{aligned} Y_1(s) &= G_1(s)U_1(s) = G_1(s)\left(V(s) - Y_2(s)\right) \\ &= G_1(s)\left(V(s) - G_2(s)Y_1(s)\right) \end{aligned}$$

whence

$$\left(I_p + G_1(s)G_2(s)\right)Y_1(s) = G_1(s)V(s)$$

and

$$Y_1(s) = \left(I_p + G_1(s)G_2(s)\right)^{-1}G_1(s)V(s)$$

where I_p is the identity matrix of size (p,p). However, it is also possible to derive the following relations

$$\begin{aligned} U_1(s) &= V(s) - G_2(s)Y_1(s) \\ &= V(s) - G_2(s)G_1(s)U_1(s) \end{aligned}$$

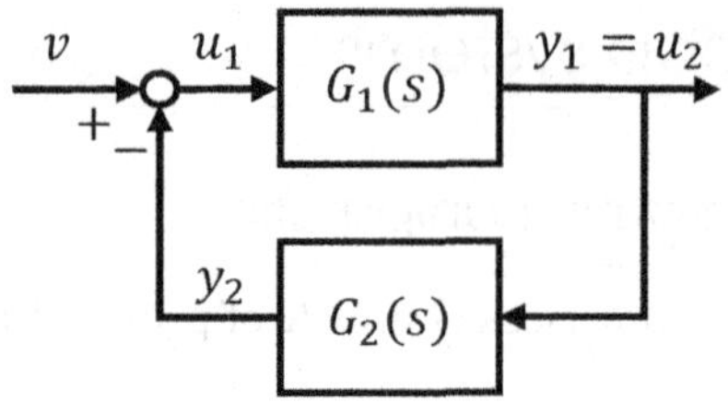

Figure 5.2: Systems in feedback configuration.

$$U_1(s) = (I_m + G_2(s)G_1(s))^{-1}V(s)$$

$$Y_1(s) = G_1(s)\,(I_m + G_2(s)G_1(s))^{-1}V(s)$$

By comparing the two previous results, and by omitting the dependence on s to simplify the notation, it follows that

$$(I_p + G_1G_2)^{-1}\,G_1 = G_1\,(I_m + G_2G_1)^{-1}$$

As a proof, note that by multiplying the previous expression from the left by $(I_p + G_1G_2)$ and from the right by $(I_m + G_2G_1)$, one obtains

$$\begin{aligned} & (I_p + G_1G_2)\,(I_p + G_1G_2)^{-1}\,G_1\,(I_m + G_2G_1) \\ = \; & (I_p + G_1G_2)\,G_1\,(I_m + G_2G_1)^{-1}\,(I_m + G_2G_1) \end{aligned}$$

$$\begin{aligned} G_1\,(I_m + G_2G_1) &= (I_p + G_1G_2)\,G_1 \\ G_1 + G_1G_2G_1 &= G_1 + G_1G_2G_1 \end{aligned}$$

Closed-loop control schemes

Consider the closed-loop system reported in Figure 5.3, where $G(s)$ and $R(s)$ are the transfer functions of the process and the controller, respectively, while the signals are vectors of suitable dimensions.

First define the loop functions

$$L(s) = G(s)R(s), \quad L_u(s) = R(s)G(s)$$

and note that they are square matrices, not necessarily of the same dimensions, also when $R(s)$ and $G(s)$ are not. By applying the previous rules of manipulation, it follows that

$$\begin{aligned} Y(s) &= (I + L(s))^{-1}\,L(s)\,(Y^o(s) - N(s)) + (I + L(s))^{-1}\,D_y(s) \\ & \quad + (I + L(s))^{-1}\,G(s)D_u(s) \\ U(s) &= (I + L_u(s))^{-1}\,R(s)\,(Y^o(s) - D_y(s) - N(s)) + (I + L_u(s))^{-1}\,D_u(s) \end{aligned}$$

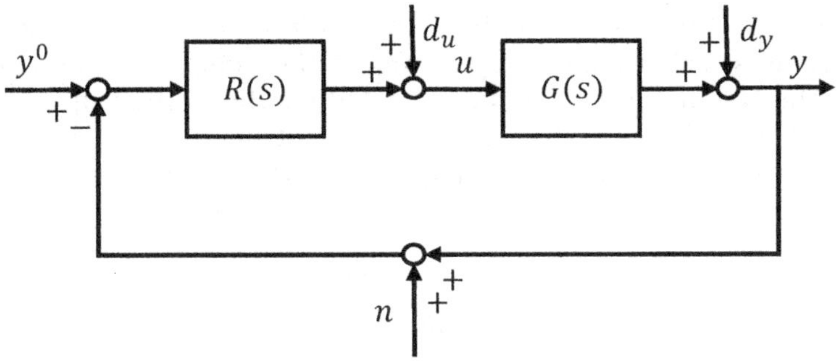

Figure 5.3: Closed-loop system.

Define also

$$\begin{array}{ll} S(s) = (I + L(s))^{-1} & \textit{sensitivity function} \\ T(s) = (I + L(s))^{-1} L(s) & \textit{complementary sensitivity function} \end{array}$$

Therefore, $Y(s)$ can be written as

$$Y(s) = T(s)\left(Y^o(s) - N(s)\right) + S(s)D_y(s) + S(s)G(s)D_u(s)$$

Moreover note that, in view of the previous considerations, the complementary sensitivity can also be written as

$$T(s) = L(s)\left(I + L(s)\right)^{-1}$$

Finally, since also in the multivariable case

$$S(s) + T(s) = I$$

it is apparent that it is not possible to design a controller guaranteeing independent specifications on $S(s)$ and $T(s)$. In the following, as in the *SISO* case, the following problems will be considered:

1. stability
2. static performance
3. dynamic performance.

5.2.2 Stability

Definition 5.2.1. *A system composed by several blocks is asymptotically stable if: (*a*) there are no hidden cancellations of unstable modes, i.e. cancellations between poles and zeros with nonnegative real part inside its blocks, and (*b*) bounded inputs applied at any point of the system produce bounded outputs at any point of the system.*

As in the *SISO* case, the stability of *MIMO* feedback systems can be studied with the Nyquist criterion, i.e. by means of the analysis of the loop transfer function. The Nyquist criterion is now stated for the closed-loop system shown in Figure 5.4, where it is assumed that the loop transfer function $L(s)$ does not contain unstable hidden modes (no pole-zero cancellations of singularities with nonnegative real part).

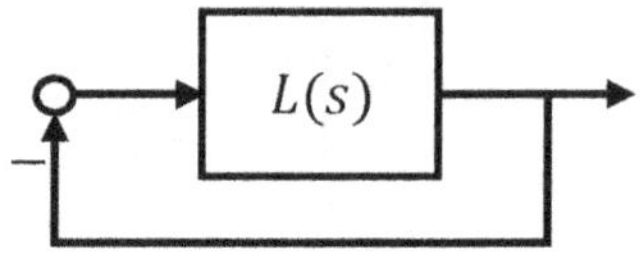

Figure 5.4: Feedback system.

Theorem 5.2.1. Let P_{ol} be the number of poles of $L(s)$ with positive real part. Then, the closed-loop system with loop transfer function $L(s)$ and negative feedback is asymptotically stable if and only if the Nyquist plot of $\det(I + L(s))$: (i) does not pass through the origin, (ii) the number of its encirclements (positive in the anticlockwise direction) around the origin is P_{ol}.

This result is conceptually very important, since it extends to the multivariable case the main stability analysis tool for *SISO* systems. However, its use for the regulator synthesis is quite difficult and hardly used in practice. Moreover, it is not possible to extend to the *MIMO* case the Bode criterion, which is de facto the most useful tool in the *SISO* case for the synthesis of the controller in the frequency domain.
For these reasons, in the next chapters a different approach will be followed: direct regulator synthesis methods guaranteeing closed-loop stability will be used. They will be developed mostly in the time domain, rather than in the frequency domain (as in the *SISO* case).

Anyway, the following results (here not proven) based on the direct inspection of the closed-loop transfer functions can be useful for the stability analysis of multivariable feedback systems.

Theorem 5.2.2. *The feedback system is internally stable if and only if the transfer functions*

$$\begin{aligned}
K_1(s) &= (I + L(s))^{-1} L(s) \\
K_2(s) &= (I + L(s))^{-1} \\
K_3(s) &= (I + L(s))^{-1} G(s) \\
K_4(s) &= (I + L_u(s))^{-1} R(s) \\
K_5(s) &= (I + L_u(s))^{-1}
\end{aligned}$$

are asymptotically stable.

The conditions of the previous Theorem guarantee that the possible presence of cancellations between the poles and the invariant zeros of $G(s)$ and $R(s)$, which could be hidden in one closed-loop transfer function, can be detected by the inspection of at least one of the remaining transfer functions. However, the following result shows that it is often sufficient to consider just one of the previous closed-loop transfer functions.

Theorem 5.2.3. *If there are not cancellations between poles of $G(s)$ and invariant zeros of $R(s)$, or viceversa, with positive or null real part, the feedback system is asymptotically stable if and only if one of the previous closed-loop transfer functions $K_i(s)$, $i = 1, 2, 3, 4, 5$, is asymptotically stable.*

Note that in the *MIMO* case the presence of cancellations must be carefully considered. In fact, even if two cascaded systems, with transfer functions $G_1(s)$ and $G_2(s)$, have a pole coinciding with an invariant zero, it can happen that the overall system, with transfer function $G_2(s)G_1(s)$ (or $G_1(s)G_2(s)$), does not have any cancellation. A cancellation between $G_1(s)$ and $G_2(s)$ exists only if the poles of $G_1(s)$ and/or of $G_2(s)$ are not included in the poles of $G_1(s)G_2(s)$ (or $G_2(s)G_1(s)$).

Example 5.2.1. Consider two systems described by

$$G_1(s) = \begin{bmatrix} \dfrac{4}{s+3} & 0 \\ 0 & \dfrac{-4(s-1)}{s} \end{bmatrix}, \quad G_2(s) = \begin{bmatrix} \dfrac{4}{s-1} & 0 \\ 0 & \dfrac{1}{s+8} \end{bmatrix}$$

The transfer function $G_1(s)$ has poles in $s = -3$ and $s = 0$ and a zero in $s = 1$, while $G_2(s)$ has poles in $s = 1$ and $s = -8$. Hence, there is a pole-zero pair in $s = 1$ (positive real). However, the cascade system described by $G_2(s)G_1(s)$ (or by $G_1(s)G_2(s)$), has transfer function

$$G_2(s)G_1(s) = \begin{bmatrix} \dfrac{16}{(s+3)(s-1)} & 0 \\ 0 & \dfrac{-4(s-1)}{s(s+8)} \end{bmatrix}$$

with poles in $s = 1$, $s = -3$, $s = -8$, $s = 0$ and a zero in $s = 1$. In this case there are no cancellations even though a pole and a zero coincide. Note that the *MIMO* system is indeed made by two independent *SISO* subsystems and, obviously, there are no cancellations between zeros of one system and poles of the other. Moreover, by means of the Nyquist criterion for the analysis of the transfer functions on the diagonal of $G_2(s)G_1(s)$, it is easy to see that the unitary negative feedback systems with loop transfer function $L(s) = G_2(s)G_1(s)$ (or $L(s) = G_1(s)G_2(s)$) is asymptotically stable.

As for the problem of robust stability, the small gain theorem, previously introduced and here recalled, can be used also in the the case of multivariable feedback systems.

Theorem 5.2.4 (Small gain theorem). *A closed-loop system made by asymptotically stable linear systems and with loop transfer function $L(s)$ is asymptotically stable if the loop gain is less than 1, that is if*

$$\|L\|_\infty < 1$$

5.2.3 Robust stability

Based on the small gain theorem, the same analysis approach already presented in Paragraphs 3.1.4 and 3.1.5 for the assessment of the robust stability of the closed-loop system in presence of additive or multiplicative uncertainties can be applied also to *MIMO* systems. The considerations already reported in Paragraphs 3.1.4 and 3.1.5 will not be repeated here. However, the analysis on robust stability in case of additive uncertainties is now further elaborated as follows. Assume that the nominal system is described by $\bar{G}(s)$, while the real system is described by the transfer function matrix

$$G(s) = \bar{G}(s) + \Delta G_a(s)$$

as shown in Figure 5.5.

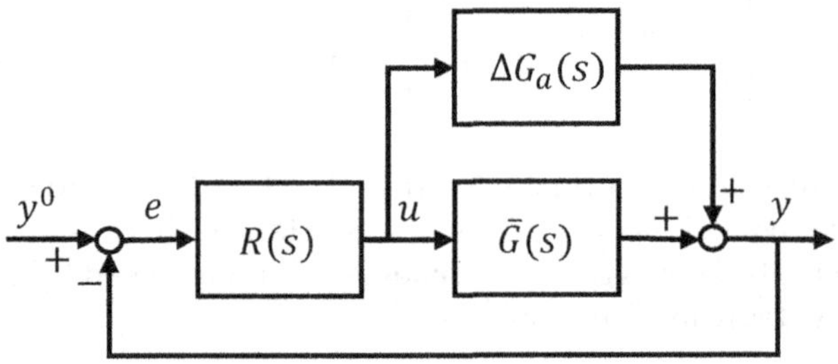

Figure 5.5: Perturbed system with additive uncertain.

It is possible to prove that the uncertainty can be factorized in the following way

$$\Delta G_a(s) = W_1(s)\bar{\Delta}G_a(s)W_2(s)$$

where $W_1(s)$ and $W_2(s)$ are known transfer function matrices without zeros and poles with positive real part and $\bar{\Delta}G_a$ is an unknown transfer function matrix such that

$$\left\|\bar{\Delta}G_a\right\|_\infty < 1$$

The previous scheme can be represented as in Figure 5.6.

It is then possible to highlight the known part of the controlled process $Q(s)$ and the uncertainty with infinity norm less than 1; see Figure 5.7 where

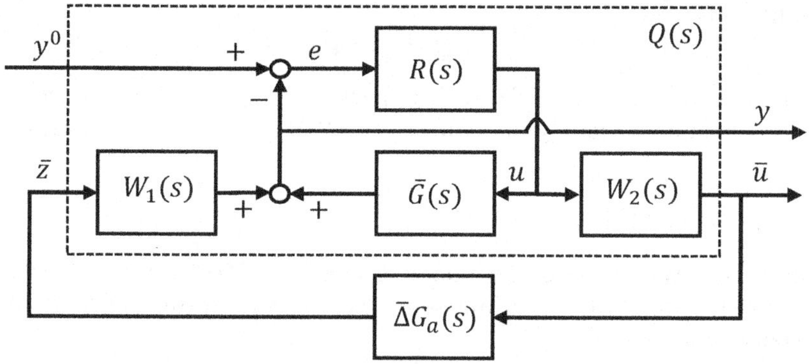

Figure 5.6: Perturbed scheme.

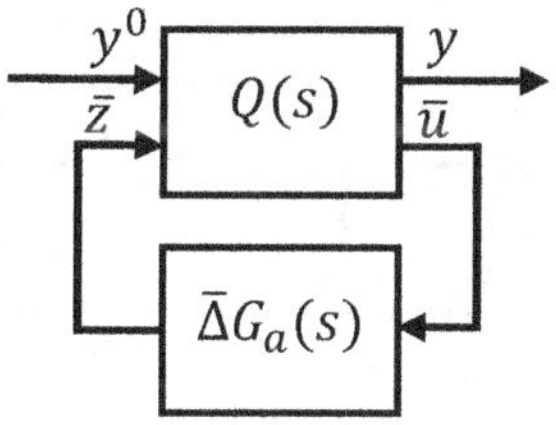

Figure 5.7: Equivalent representation of the perturbed system.

$$\begin{bmatrix} Y(s) \\ \bar{U}(s) \end{bmatrix} = Q(s) \begin{bmatrix} Y^0(s) \\ \bar{Z}(s) \end{bmatrix} = \begin{bmatrix} Q_{11}(s) & Q_{12}(s) \\ Q_{21}(s) & Q_{22}(s) \end{bmatrix} \begin{bmatrix} Y^0(s) \\ \bar{Z}(s) \end{bmatrix}$$

and, omitting the dependence on s,

$$\begin{aligned} Q_{11} &= (I + \bar{G}R)^{-1}\bar{G}R \\ Q_{12} &= (I + \bar{G}R)^{-1}W_1 \\ Q_{21} &= W_2(I + R\bar{G})^{-1}R \\ Q_{22} &= -W_2(I + R\bar{G})^{-1}RW_1 \end{aligned}$$

By Theorem 5.2.4, it follows that if $\bar{\Delta}G_a$ and the nominal system (Q_{22}) are stable, a sufficient condition for robust stability is

$$\left\| Q_{22}\bar{\Delta}G_a \right\|_\infty < 1$$

Since, by assumption,

$$\left\| \bar{\Delta}G_a \right\|_\infty < 1$$

the condition becomes

$$\|Q_{22}\|_\infty < 1$$

In view of the definition of Q_{22}, it is possible to verify that this condition is similar to the one previously introduced for *SISO* systems in Paragraph 3.1.4.

5.2.4 Static performance

The analysis of the static performance of feedback *MIMO* systems reported in this section is aimed at extending some well known results of *SISO* systems, here briefly summarized.

Consider the *SISO* system reported in Figure 5.8, where

$$L(s) = \frac{\mu}{s^g} \frac{\prod(1+\tau_i s)}{\prod(1+T_j s)}$$

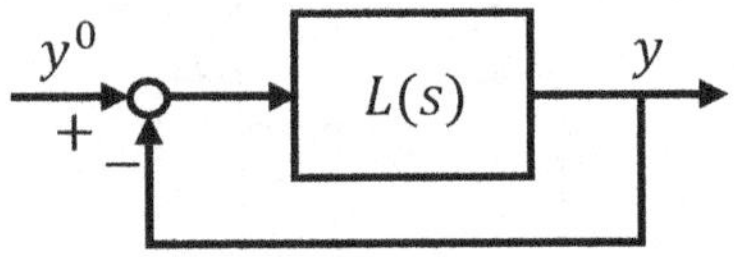

Figure 5.8: Feedback *SISO* system.

Assuming that the feedback system is asymptotically stable, given a constant set-point with Laplace transform $Y^0(s) = \frac{A}{s}$, the output asymptotically tends to

$$y_\infty = \lim_{t\to\infty} y(t) = \lim_{s\to 0} s \frac{L(s)}{1+L(s)} \frac{A}{s} = \begin{cases} \frac{\mu}{1+\mu} A & g = 0 \\ A & g > 0 \end{cases}$$

Hence, $e_\infty = y^0 - y_\infty$ is null for $g > 0$, while for $g = 0$ it is decreasing for an increasing static loop gain μ. In particular, assuming that $R(s)$ has an integral action, i.e. $\frac{1}{s}R'(s)$, the result for $g > 0$ can be interpreted as shown in Figure 5.9: since the feedback system is asymptotically stable, for constant set-points or disturbances on the output, all the signals in the loop tend to a constant value. However, the only admissible constant value for the integrator input, that is the error (or a signal proportional to the error), is zero. In fact, for any other constant value the output of the integrator would be a ramp. Observe also that the integral action guarantees *robust* zero error regulation, i.e. also for small process parameter perturbations or model errors, provided that the closed-loop stability is maintained. In fact, the previous considerations guarantee that at the steady state the input of the integrator is always null.

In view of these considerations, in order to guarantee an asymptotically small, or null, error it is necessary that $|L(j\omega)|$ is "large" for "small" value of ω, as illustrated in Figure 5.10. Note also that the design of the regulator $R(s)$ is usually

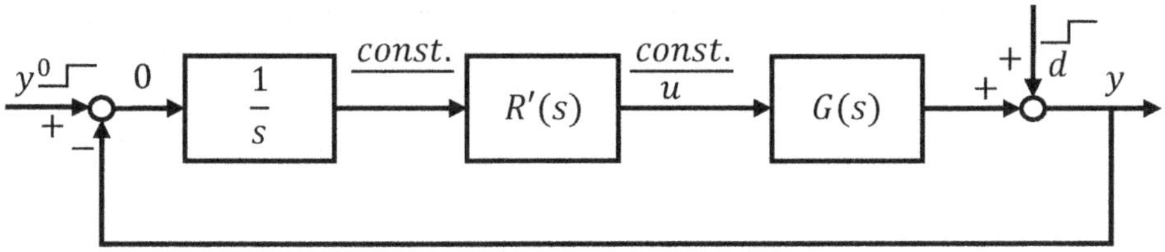

Figure 5.9: Steady state loop signals.

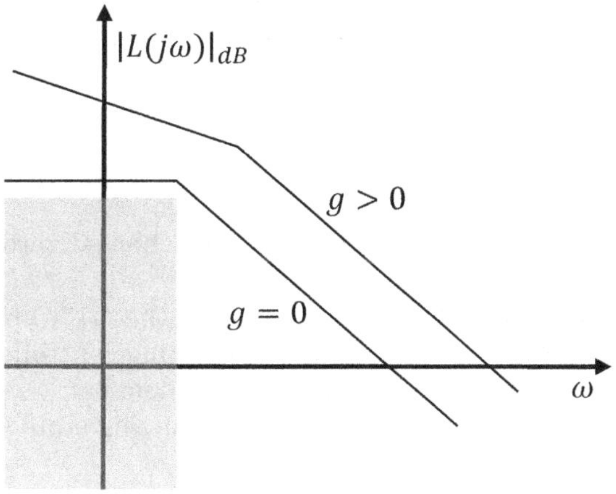

Figure 5.10: Static performance constraints for *SISO* systems.

performed by satisfying the static conditions with an integral action in the regulator transfer function (term $\frac{1}{s}$), and then by considering dynamic requirements in the design of the transfer function $R'(s)$. Of course, the inclusion of an integral action in the regulator is allowed if the process transfer function $G(s)$ does not have zeros at the origin, that is derivative actions, so that there are not cancellations of non asymptotically stable singularities. On the other hand, if $G(s)$ has one or more derivative actions, its steady state output can only be null, since the derivative of a constant signal is zero.

Schemes with integral action on the error

In close analogy with the previous discussion for *SISO* systems, also in the *MIMO* case it is possible to guarantee a null steady state error by forcing an integral action on each component of the error signal, as shown in Figure 5.11. If the closed-loop system is asymptotically stable for a suitable choice of $R'(s)$, considerations analogous to those previously developed for *SISO* systems, allow to show that, for constant set-points, the unique feasible steady state condition

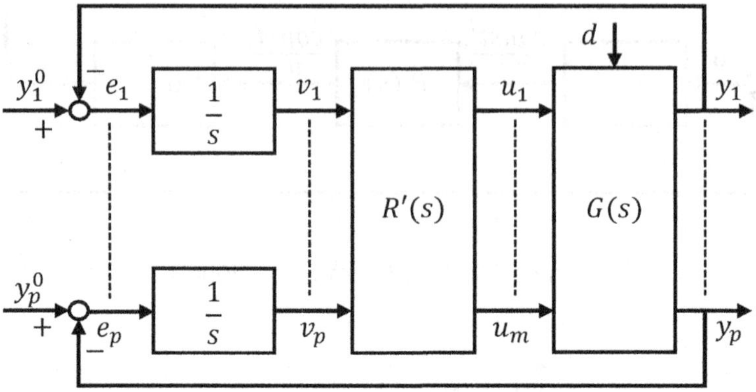

Figure 5.11: Scheme with integral action on the error for *MIMO* systems.

is such that $e_i = 0$, $i = 1, ..., p$. Moreover, the achieved zero error regulation is "robust" in the sense specified above.

With this scheme, the static requirements on the error are automatically verified, but the problem of synthesizing the stabilizing controller $R'(s)$ is still open. To analyze the conditions under which this problem can be solved, assume that the system under control is described by the following state space model

$$\begin{aligned} \dot{x}(t) &= Ax(t) + Bu(t) + Md \\ y(t) &= Cx(t) + Nd \end{aligned}$$

where $x \in R^n$, $u \in R^m$, $y \in R^p$, and $d \in R^r$ is a constant disturbance acting on the state and on the output. For this system, it is required that, at the steady state, $y = y^0$ for any constant y^0 and d.

Denoting with barred symbols the steady state variables, to guarantee zero steady state error the following relations must hold

$$\begin{aligned} 0 &= A\bar{x} + B\bar{u} + Md \\ y^0 &= C\bar{x} + Nd \end{aligned}$$

that is

$$\begin{bmatrix} A & B \\ C & 0 \end{bmatrix} \begin{bmatrix} \bar{x} \\ \bar{u} \end{bmatrix} = \begin{bmatrix} 0 & -M \\ I & -N \end{bmatrix} \begin{bmatrix} y^0 \\ d \end{bmatrix}$$

Now note that

$$\Sigma = \begin{bmatrix} A & B \\ C & 0 \end{bmatrix} \in R^{n+p,n+m}$$

Therefore, the conditions $p \leq m$ and $rank(\Sigma) = n+p$ must hold to guarantee that there exists at least a pair $\bar{x}$, $\bar{u}$ satisfying the previous relation for any constant y^0 and d. It is easy to interpret the meaning of the condition $p \leq m$: if it is not fulfilled, there does not exist a sufficient number of control variables to force arbitrary values to the outputs y. Recalling the definition of invariant zeros, the condition $rank(\Sigma) = n + p$ is equivalent to require that the system under control does not have invariant zeros in $s = 0$, i.e. derivative actions. This is necessary to exclude cancellations between the poles of the integrators and process derivative actions. To derive in a more formal way this condition, observe that the controller $R'(s)$ must be synthesized to stabilize the enlarged system made by the process and the integrators. The integrators in state variables are described by

$$\begin{aligned} \dot{v}(t) &= e(t) \\ &= y^0 - y(t) \\ &= y^0 - Cx(t) - Nd \end{aligned}$$

Hence, the enlarged system made by the process and the integrators, with manipulable input u and measured output v, is

$$\begin{aligned} \begin{bmatrix} \dot{x}(t) \\ \dot{v}(t) \end{bmatrix} &= \begin{bmatrix} A & 0 \\ -C & 0 \end{bmatrix} \begin{bmatrix} x(t) \\ v(t) \end{bmatrix} + \begin{bmatrix} B \\ 0 \end{bmatrix} u(t) + \begin{bmatrix} 0 & M \\ I & -N \end{bmatrix} \begin{bmatrix} y^0 \\ d \end{bmatrix} \\ v(t) &= \begin{bmatrix} 0 & I \end{bmatrix} \begin{bmatrix} x(t) \\ v(t) \end{bmatrix} \end{aligned}$$

Defining

$$\bar{A} = \begin{bmatrix} A & 0 \\ -C & 0 \end{bmatrix}, \quad \bar{B} = \begin{bmatrix} B \\ 0 \end{bmatrix}, \quad \bar{C} = \begin{bmatrix} 0 & I \end{bmatrix}$$

in order to synthesize the stabilizing regulator $R'(s)$, the pair $(\bar{A}, \bar{B})$ must be reachable (or stabilizable) and the pair $(\bar{A}, \bar{C})$ must be observable (or detectable). These two conditions are analyzed in the following.

Observability of$(\bar{A}, \bar{C})$

The observability matrix is

$$\bar{M}_o = \begin{bmatrix} \bar{C} \\ \bar{C}\bar{A} \\ \bar{C}\bar{A}^2 \\ \vdots \end{bmatrix} = \begin{bmatrix} 0 & I \\ -C & 0 \\ -CA & 0 \\ \vdots & \vdots \end{bmatrix}$$

Clearly this matrix is full rank if the pair (A, C) is observable.

Reachability of$(\bar{A}, \bar{B})$

The reachability matrix is

$$\begin{aligned} \bar{M}_r &= \begin{bmatrix} \bar{B} & \bar{A}\bar{B} & \bar{A}^2\bar{B} & \dots \end{bmatrix} = \\ &= \begin{bmatrix} B & AB & A^2B & A^3B & \dots \\ 0 & -CB & -CAB & -CA^2B & \dots \end{bmatrix} \\ &= \begin{bmatrix} A & B \\ -C & 0 \end{bmatrix} \begin{bmatrix} 0 & B & AB & A^2B & \dots \\ I & 0 & 0 & 0 & \dots \end{bmatrix} \end{aligned}$$

In view of the reachability of the pair (A, B) and of the assumption that the $rank(\Sigma) = n + p$ $(p \leq m)$, since the rank of

$$\Sigma' := \begin{bmatrix} A & B \\ -C & 0 \end{bmatrix}$$

is equal to the one of Σ, it follows that $rank(\bar{M}_r) = n + p$ so that the pair $(\bar{A}, \bar{B})$ is reachable.

As already anticipated, if $rank(\Sigma) < n + p$, the process has invariant zeros in $s = 0$, that is it has derivative actions which cancel with the poles of the integrators, causing loss of reachability.

5.2.5 Dynamic performance

Going back to *MIMO* systems, as it has been previously shown, in the Laplace transform domain the relation between the set-point and the output is

$$Y(s) = T(s)Y^o(s)$$

where $T(s)$ is the complementary sensitivity function. In order to have a perfect tracking of the reference signal, one should require that $T(j\omega) \simeq I$, at least in the frequency band where the reference has a significant harmonic content. This is equivalent to require that

$$T(j\omega) = I - S(j\omega) = I - (I + L(j\omega))^{-1} \simeq I$$

or also

$$S(j\omega) \simeq 0$$

that is

$$\frac{\left\| (I + L(j\omega))^{-1} x \right\|_2}{\|x\|_2} \ll 1, \quad \forall x$$

This is guaranteed if

$$\bar{\sigma}\left[(I + L(j\omega))^{-1} \right] \ll 1$$

that is (see Appendix A)

$$\bar{\sigma}\left[(I + L(j\omega))^{-1} \right] = \frac{1}{\underline{\sigma}\left[(I + L(j\omega)) \right]} \ll 1$$

Recalling that

$$max(0, \underline{\sigma}(Q) - 1) \leq \underline{\sigma}(Q + I) \leq \underline{\sigma}(Q) + 1$$

it is possible to write

$$\frac{1}{\underline{\sigma}\left[(I + L(j\omega)) \right]} \leq \frac{1}{\underline{\sigma}\left[(L(j\omega)) \right] - 1}$$

and, if $\underline{\sigma}\left[(L(j\omega))\right] \gg 1$,

$$\bar{\sigma}\left[(I + L(j\omega))^{-1}\right] \leq \frac{1}{\underline{\sigma}\left[(L(j\omega))\right]}$$

In conclusion, with some approximations, the requirement $T(j\omega) \simeq I$ becomes

$$\underline{\sigma}\left(L(j\omega)\right) \gg 1$$

This condition is analogous to the one obtained for the static gain of *SISO* systems, when the considered frequency band is defined for "small"values of ω. The obtained condition is schematically illustrated in Figure 5.12.

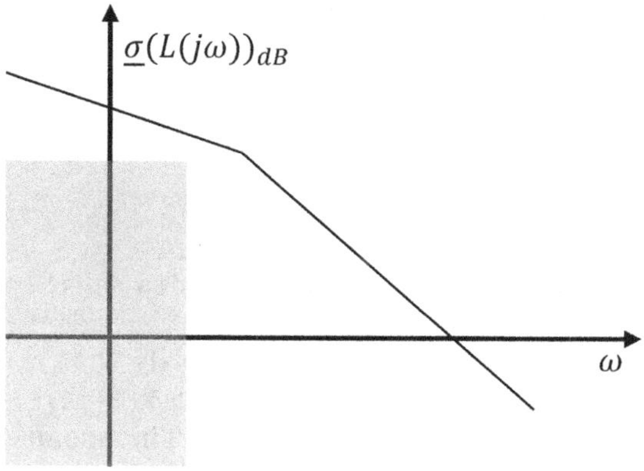

Figure 5.12: Constraints on the smallest singular value for dynamic performance of *MIMO* systems.

As for the rejection of the measurement noise, recall that

$$Y(s) = T(s)\left(Y^o(s) - N(s)\right) + S(s)D_y(s)$$

Then, one can require that $T(j\omega)$ "is small" where $N(j\omega)$ has significant harmonic components, typically at high frequency.

This requirement can be elaborated as follows. First write again

$$T(s) = I - S(s) = I - (I + L(s))^{-1}$$

and, in view of the matrix inversion lemma,

$$T(s) = I - (I + L(s))^{-1} = (I + L(s)^{-1})^{-1}$$

Therefore,

$$\bar{\sigma}(T(j\omega)) = \bar{\sigma}((I + L(j\omega)^{-1})^{-1})$$

and

$$\bar{\sigma}(T(j\omega)) = \bar{\sigma}((I + L(j\omega)^{-1})^{-1}) = \frac{1}{\underline{\sigma}\left(I + L(j\omega)^{-1}\right)} \leq \frac{1}{\underline{\sigma}\left(L(j\omega)^{-1}\right) - 1}$$

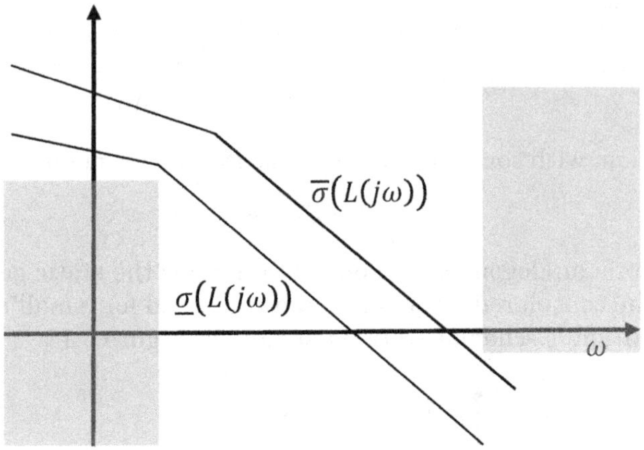

Figure 5.13: Dynamic constraints on the minimum and maximum singular values.

so that, in order to have $T(j\omega)$ "small"one has to select $\underline{\sigma}(L(j\omega)^{-1})$ "big", i.e. $\bar{\sigma}(L(j\omega))$ "small".
With similar considerations it is possible to verify that the need to reject the disturbance $D_y(s)$ is equivalent to require that $\bar{\sigma}\,(S(j\omega))$ is "small" where the spectrum of D_y lies, typically at low frequency. This means that $\underline{\sigma}(L(j\omega))$ must be "big" at low values of ω. All these considerations are summarized in Figure 5.13.

As for the performance associated with the control variable, recall that

$$U(s) = (I + R(s)G(s))^{-1}\,R(s)\,(Y^o(s) - D_y(s) - N(s))$$

For the control sensitivity function (omitting the explicit dependence on ω), the following result holds

$$\bar{\sigma}\left((I+RG)^{-1}R\right) \leq \frac{\bar{\sigma}\,(R)}{\underline{\sigma}\,(I+RG)} \approx \begin{cases} \dfrac{\bar{\sigma}\,(R)}{\underline{\sigma}\,(RG)} & \text{if} \quad \underline{\sigma}\,(RG) \gg 1 \\ \bar{\sigma}\,(R) & \text{if} \quad \underline{\sigma}\,(RG) \ll 1 \end{cases}$$

This result is similar to the one that can be stated for *SISO* systems, and allows one to evaluate the effects of the exogenous signals on the control variable produced by the feedback system.

6
Pole placement control

In this chapter, the pole placement control technique is developed for continuous-time linear systems. First, the case of systems with measurable state is considered. Then, an output feedback control law is developed by the use of state observers.

6.1 Measurable state

Consider the continuous-time system

$$\begin{aligned} \dot{x}(t) &= Ax(t) + Bu(t), \quad x \in R^n, \quad u \in R^m \\ y(t) &= Cx(t) + Du(t), \quad y \in R^p \end{aligned}$$

where the state x is assumed to be measurable, and the following algebraic control law

$$u(t) = -Kx(t) + v(t), \quad K \in R^{m,n}, \quad v \in R^m$$

By combining the state equation and the control law, the closed-loop system is

$$\dot{x}(t) = (A - BK)x(t) + Bv(t)$$

Then, the pole placement problem consists of computing a matrix K such that the eigenvalues of $A - BK$ can be arbitrarily assigned.

Note that the signal v does not play any role in the problem solution, but it can be used to satisfy other requirements. For example, the previous control law can be designed to stabilize an unstable system by assigning the eigenvalues of $A - BK$, while the input v can be used to close an outer control loop to meet static or dynamic specifications.

The following result holds.

Theorem 6.1.1. *A necessary and sufficient condition for the solution of the pole placement problem (with measurable state) is that the pair (A, B) is reachable.*

Proof. Necessity is now proven, while sufficiency will be shown in the next subsection by means of a constructive procedure. To this end, recall that if (A, B) is not reachable, then there exists a transformation $\hat{x} = Tx$ such that the transformed system takes the form

$$\begin{bmatrix} \dot{\hat{x}}_r(t) \\ \dot{\hat{x}}_{nr}(t) \end{bmatrix} = \begin{bmatrix} \hat{A}_r & \hat{A}_x \\ 0 & \hat{A}_{nr} \end{bmatrix} \begin{bmatrix} \hat{x}_r(t) \\ \hat{x}_{nr}(t) \end{bmatrix} + \begin{bmatrix} \hat{B}_r \\ 0 \end{bmatrix} u(t)$$

where $\hat{x}_r \in R^{\nu_r}$, $\nu_r = rankM_r$, $M_r = \begin{bmatrix} B & AB & ... & A^{n-1}B \end{bmatrix}$.

With reference to this transformed system, consider the following control law

$$u(t) = -\hat{K}\hat{x}(t) = -\begin{bmatrix} \hat{K}_r & \hat{K}_{nr} \end{bmatrix} \begin{bmatrix} \hat{x}_r(t) \\ \hat{x}_{nr}(t) \end{bmatrix}, \quad \hat{K}_r \in R^{m,\nu_r}$$

It follows that

$$\hat{A} - \hat{B}\hat{K} = \begin{bmatrix} \hat{A}_r - \hat{B}_r\hat{K}_r & \hat{A}_x - \hat{B}_r\hat{K}_{nr} \\ 0 & \hat{A}_{nr} \end{bmatrix}$$

Therefore, the eigenvalues of the closed-loop system are those of $(\hat{A}_r - \hat{B}_r\hat{K}_r)$ and of the unreachable part $\hat{A}_{nr}$, which cannot be modified by any choice of $\hat{K}$. Since the eigenvalues of $\hat{A}$ and A are the same, it is apparent the eigenvalues of the unreachable part of A cannot be modified and the pole placement problem cannot be solved.

Remark. If the system is only stabilizable, the unreachable part has stable eigenvalues, i.e. the eigenvalues of $\hat{A}_{nr}$ have negative real part. In this case, the pole placement algorithm described in the following can be applied to the reachable part only, defined by the pair $(\hat{A}_r, \hat{B}_r)$.

In the following, it will always be assumed that the system is reachable.

6.1.1 Systems with scalar input ($m = 1$)

The pole placement problem is first solved for systems with $m = 1$. In this case, there exists a state transformation $\tilde{x} = Tx$ such that the transformed system is in the controllable (or reachable) canonical form

$$\dot{\tilde{x}}(t) = \tilde{A}\tilde{x}(t) + \tilde{B}u(t)$$

where

$$\tilde{A} = TAT^{-1} = \begin{bmatrix} 0 & 1 & 0 & ... & 0 \\ 0 & 0 & 1 & ... & 0 \\ ... & ... & ... & ... & ... \\ 0 & 0 & 0 & ... & 1 \\ -a_0 & -a_1 & -a_2 & ... & -a_{n-1} \end{bmatrix}, \quad \tilde{B} = TB = \begin{bmatrix} 0 \\ 0 \\ ... \\ 0 \\ 1 \end{bmatrix}$$

and

$$\det(sI - A) = \det\left(sI - \tilde{A}\right) = s^n + a_{n-1}s^{n-1} + ... + a_1 s + a_0$$

Remark. The transformation matrix T can be determined as follows. First, define the reachability matrices of the original system and of the system in the controllable canonical form.

$$\begin{aligned} M_r &= \begin{bmatrix} B & AB & ... & A^{n-1}B \end{bmatrix} \\ \tilde{M}_r &= \begin{bmatrix} \tilde{B} & \tilde{A}\tilde{B} & ... & \tilde{A}^{n-1}\tilde{B} \end{bmatrix} \\ &= \begin{bmatrix} TB & TAT^{-1}TB & ... & (TAT^{-1}TAT^{-1}...TB) \end{bmatrix} \\ &= T\begin{bmatrix} B & AB & ... & A^{n-1}B \end{bmatrix} = TM_r \end{aligned}$$

It follows that

$$T = \tilde{M}_r M_r^{-1}$$

Note that the matrices $\tilde{A}$ and $\tilde{M}_r$, can be built once the characteristic polynomial of A is known. Moreover, the inverse of M_r exists in view of the reachability assumption. □

In the transformed variables, the control law (the additional input v is now omitted) can be written as

$$u(t) = -Kx(t) = -KT^{-1}\tilde{x}(t) = -\tilde{K}\tilde{x}(t)$$

which leads to the closed-loop system

$$\dot{\tilde{x}}(t) = (\tilde{A} - \tilde{B}\tilde{K})\tilde{x}(t)$$

Moreover

$$\tilde{A} - \tilde{B}\tilde{K} = TAT^{-1} - TBKT^{-1} = T(A - BK)T^{-1}$$

then, the eigenvalues of $\tilde{A} - \tilde{B}\tilde{K}$ and $A - BK$ are the same, and the pole placement problem is solved provided that a matrix $\tilde{K}$ can be found such that the eigenvalues of $\tilde{A} - \tilde{B}\tilde{K}$ can be arbitrarily assigned.

To this end, consider the matrix

$$\tilde{K} = \begin{bmatrix} \tilde{k}_0 & \tilde{k}_1 & \dots & \tilde{k}_{n-1} \end{bmatrix}$$

In view of the structure of $\tilde{A}$ and $\tilde{B}$ it follows that

$$\tilde{A} - \tilde{B}\tilde{K} = \begin{bmatrix} 0 & 1 & 0 & \dots & 0 \\ 0 & 0 & 1 & \dots & 0 \\ \dots & \dots & \dots & \dots & \dots \\ 0 & 0 & 0 & \dots & 1 \\ -a_0 - \tilde{k}_0 & -a_1 - \tilde{k}_1 & -a_2 - \tilde{k}_2 & \dots & -a_{n-1} - \tilde{k}_{n-1} \end{bmatrix}$$

and

$$P(s) = \det(sI - (\tilde{A} - \tilde{B}\tilde{K})) = s^n + (a_{n-1} + \tilde{k}_{n-1})s^{n-1} + \dots + (a_1 + \tilde{k}_1)s + (a_0 + \tilde{k}_0)$$

Now, in order to impose that the eigenvalues of the closed-loop system are real or complex conjugate $\{-\bar{p}_1, -\bar{p}_2, \dots, -\bar{p}_n\}$, its characteristic polynomial must be

$$P(s) = (s + \bar{p}_1)(s + \bar{p}_2)\dots(s + \bar{p}_n) = s^n + p_{n-1}s^{n-1} + \dots + p_1 s + p_0$$

It is then sufficient to select

$$a_i + \tilde{k}_i = p_i, \quad i = 0, \dots, n-1$$

that is

$$\tilde{k}_i = -a_i + p_i, \quad i = 0, \dots, n-1$$

Once $\tilde{K}$ has been computed, it is easy to obtain K as follows

$$K = \tilde{K}T$$

In summary, the problem initially stated can be solved with the following algorithm.

Given A, B and the desired characteristic polynomial $P(s) = s^n + p_{n-1}s^{n-1} + \ldots + p_1 s + p_0$:

1. compute the characteristic polynomial of A, that is its coefficients a_i;
2. build the matrices $\tilde{A}$ and $\tilde{B}$;
3. compute M_r, $\tilde{M}_r$ and $T = \tilde{M}_r M_r^{-1}$;
4. set $\tilde{K} = \begin{bmatrix} -a_0 + p_0 & -a_1 + p_1 & \ldots & -a_{n-1} + p_{n-1} \end{bmatrix}$;
5. compute $K = \tilde{K}T$.

However, the computation of the controllable canonical form is not really necessary, as shown below.

The Ackermann's formula

Evaluate the required characteristic polynomial $P(s)$ in $\tilde{A}$

$$P(\tilde{A}) = \tilde{A}^n + p_{n-1}\tilde{A}^{n-1} + \ldots + p_1\tilde{A} + p_0 I$$

In view of the Cayley-Hamilton theorem, the following expression holds

$$\tilde{A}^n = -a_{n-1}\tilde{A}^{n-1} - \ldots - a_1\tilde{A} - a_0 I$$

By substituting this expression into the one of $P(\tilde{A})$ one obtains

$$P(\tilde{A}) = (p_{n-1} - a_{n-1})\tilde{A}^{n-1} + \ldots + (p_1 - a_1)\tilde{A} + (p_0 - a_0)I$$

In order to study the structure of the matrices $\tilde{A}^i$, $i = 0, \ldots, n-1$, consider the simple case of a third order system. Simple computations show that

$$\tilde{A}^0 = \begin{bmatrix} \mathbf{1} & 0 & 0 \\ 0 & 1 & 0 \\ 0 & 0 & 1 \end{bmatrix}, \quad \tilde{A}^1 = \begin{bmatrix} 0 & \mathbf{1} & 0 \\ 0 & 0 & 1 \\ * & * & * \end{bmatrix}, \quad \tilde{A}^2 = \begin{bmatrix} 0 & 0 & \mathbf{1} \\ * & * & * \\ * & * & * \end{bmatrix}$$

where the elements $*$ are "don't care" terms. Note (bold terms) that the 1 in the first row of $\tilde{A}^0$ moves along the columns (from the first to the last one) for increasing powers of $\tilde{A}$. It then follows that

$$P(\tilde{A}) = \begin{bmatrix} 0 & 0 & \mathbf{p_2 - a_2} \\ * & * & * \\ * & * & * \end{bmatrix} + \begin{bmatrix} 0 & \mathbf{p_1 - a_1} & 0 \\ * & * & * \\ * & * & * \end{bmatrix} + \begin{bmatrix} \mathbf{p_0 - a_0} & 0 & 0 \\ * & * & * \\ * & * & * \end{bmatrix}$$

Hence

$$\tilde{K} = \begin{bmatrix} p_0 - a_0 & p_1 - a_1 & p_2 - a_2 \end{bmatrix} = \begin{bmatrix} 1 & 0 & 0 \end{bmatrix} P(\tilde{A})$$

If the system is of order n, one has

$$\begin{aligned}\tilde{K} &= \begin{bmatrix} -a_0+p_0 & -a_1+p_1 & \dots & -a_{n-2}+p_{n-2} & -a_{n-1}+p_{n-1} \end{bmatrix} \\ &= \begin{bmatrix} 1 & 0 & \dots & 0 & 0 \end{bmatrix} P(\tilde{A})\end{aligned}$$

and, since $K = \tilde{K}T$,

$$K = \begin{bmatrix} 1 & 0 & \dots & 0 & 0 \end{bmatrix} P(\tilde{A})T$$

Recall that $\tilde{A} = TAT^{-1}$ and $P(\tilde{A}) = P(TAT^{-1}) = TP(A)T^{-1}$, therefore, also by recalling that $T = \tilde{M}_r M_r^{-1}$,

$$K = \begin{bmatrix} 1 & 0 & \dots & 0 & 0 \end{bmatrix} TP(A) = \begin{bmatrix} 1 & 0 & \dots & 0 & 0 \end{bmatrix} \tilde{M}_r M_r^{-1} P(A)$$

Now consider the structure of $\tilde{M}_r$ and, for example, the case of a third order system

$$\tilde{M}_r = \begin{bmatrix} \tilde{B} & \tilde{A}\tilde{B} & \tilde{A}^2\tilde{B} \end{bmatrix}$$

where

$$\begin{aligned}\tilde{B} &= \begin{bmatrix} 0 \\ 0 \\ \mathbf{1} \end{bmatrix}, \quad \tilde{A}\tilde{B} = \begin{bmatrix} 0 & 1 & 0 \\ 0 & 0 & 1 \\ * & * & * \end{bmatrix} \begin{bmatrix} 0 \\ 0 \\ 1 \end{bmatrix} = \begin{bmatrix} 0 \\ \mathbf{1} \\ * \end{bmatrix} \\ \tilde{A}^2\tilde{B} &= \begin{bmatrix} 0 & 1 & 0 \\ 0 & 0 & 1 \\ * & * & * \end{bmatrix} \begin{bmatrix} 0 \\ 1 \\ * \end{bmatrix} = \begin{bmatrix} \mathbf{1} \\ 1 \\ * \end{bmatrix}\end{aligned}$$

and

$$\tilde{M}_r = \begin{bmatrix} \mathbf{0} & \mathbf{0} & \mathbf{1} \\ 0 & 1 & * \\ 1 & * & * \end{bmatrix}$$

In general, also in the case of higher order systems, one has

$$\tilde{M}_r = \begin{bmatrix} \mathbf{0} & \mathbf{0} & \dots & \mathbf{0} & \mathbf{1} \\ 0 & 0 & \dots & 1 & * \\ \dots & \dots & \dots & \dots & \dots \\ 0 & 1 & \dots & * & * \\ 1 & * & \dots & * & * \end{bmatrix}$$

and

$$K = \begin{bmatrix} 1 & 0 & \dots & 0 & 0 \end{bmatrix} \tilde{M}_r M_r^{-1} P(A) = \begin{bmatrix} 0 & 0 & \dots & 0 & 1 \end{bmatrix} M_r^{-1} P(A)$$

This is the so-called Ackermann's formula, which allows one to compute the matrix K without the need to preliminarily transform the system in controllable canonical form.

In conclusion, note that for single-input systems it is possible to solve the pole placement problem with a static (algebraic) regulator acting on the system state.

6.1.2 Systems with multiple inputs ($m > 1$)

When $m > 1$ the pole placement problem can have many solutions. In any case, the fundamental assumption that the pair (A, B) is reachable is required.

Example 6.1.1. Consider the system with $n = 3$, $m = 2$ and described by the matrices

$$A = \begin{bmatrix} 0 & 1 & 0 \\ 0 & 0 & 1 \\ 1 & 1 & 0 \end{bmatrix}, \quad B = \begin{bmatrix} b_1 & b_2 \end{bmatrix} = \begin{bmatrix} 1 & 0 \\ 0 & 1 \\ 0 & 0 \end{bmatrix}$$

It is easy to see that the system is reachable both from the input u_1, and from the input u_2. In fact,

$$M_{r1} = \begin{bmatrix} b_1 & Ab_1 & A^2 b_1 \end{bmatrix} = \begin{bmatrix} 1 & 0 & 0 \\ 0 & 0 & 1 \\ 0 & 1 & 0 \end{bmatrix}, \quad \det M_{r1} = -1$$

and

$$M_{r2} = \begin{bmatrix} b_2 & Ab_2 & A^2 b_2 \end{bmatrix} = \begin{bmatrix} 0 & 1 & 0 \\ 1 & 0 & 1 \\ 0 & 1 & 1 \end{bmatrix}, \quad \det M_{r2} = -1$$

It is then possible to assign the closed-loop eigenvalues with the control law

$$\begin{aligned} u_1(t) &= -K_1 x(t) + v_1(t) \\ u_2(t) &= v_2(t) \end{aligned}$$

where only the first control variable (u_1) is used and the gain matrix K_1 is computed with the Ackermann's formula applied to the pair (A, b_1). However, it is also possible to solve the problem with the control law

$$\begin{aligned} u_1(t) &= v_1(t) \\ u_2(t) &= -K_2 x(t) + v_2(t) \end{aligned}$$

where only the second control variable (u_2) is used and the gain matrix K_2 is designed for the pair(A, b_2).

In both these cases, the control law can be written in the matrix form

$$u(t) = -Kx(t) + v(t)$$

where

$$K = \begin{bmatrix} K_1 \\ 0 \end{bmatrix} \text{ (first solution)}$$

or

$$K = \begin{bmatrix} 0 \\ K_2 \end{bmatrix} \text{ (second solution)}$$

If the system is not reachable from only one input, as in the previous example, but the pair (A, B) is reachable, different solutions can be found and many algorithms have been developed to compute K. For instance, it is possible to compute the feedback gain K guaranteeing "robustness", i.e. such that the closed-loop eigenvalues are insensitive (to the maximum possible extent) to perturbations of the matrices A and B. These methods will not be further examined here.

6.2 Nonmeasurable state

When the state is not measurable, the solution of the pole assignment problem is obtained in two steps:

1. first a *state observer* is built. The observer is a system which, based on the input and output measurements, computes an estimate $\hat{x}$ of the system's state x. The observer introduced in the following is a linear time-invariant system with prescribed eigenvalues;
2. the estimated state $\hat{x}$ is used in the pole assignment control law in place of the real state x.

The corresponding control system is shown in Figure 6.1.

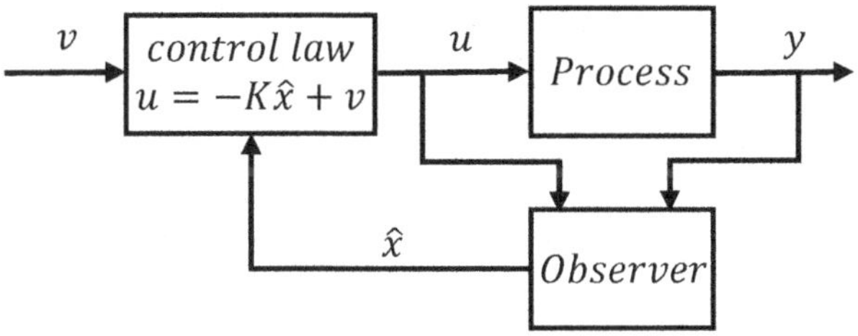

Figure 6.1: Control scheme with state observer.

6.2.1 State observers

Consider the system

$$\begin{aligned} \dot{x}(t) &= Ax(t) + Bu(t), \quad x \in R^n, \quad u \in R^m \\ y(t) &= Cx(t) + Du(t), \quad y \in R^p \end{aligned}$$

and the state observer described by

$$\dot{\hat{x}}(t) = A\hat{x}(t) + Bu(t) + L\left[y(t) - C\hat{x}(t) - Du(t)\right]$$

where $L \in R^{n,p}$ is a gain matrix to be suitably selected. Note that the observer mimics the system's dynamics plus a corrective term given by $Le_y(t)$, where $e_y(t) = y(t) - C\hat{x}(t) - Du(t)$, is the output estimation error. If $\hat{x}(t) = x(t)$, then $e_y(t) = 0$.

Denoting by $e(t) = x(t) - \hat{x}(t)$ the state estimation error, in view of the previous equations one has

$$\begin{aligned} \dot{e}(t) &= \dot{x}(t) - \dot{\hat{x}}(t) \\ &= Ax(t) + Bu(t) - A\hat{x}(t) - Bu(t) - L\left[y(t) - C\hat{x}(t) - Du(t)\right] \\ &= A(x(t) - \hat{x}(t)) - LC(x(t) - \hat{x}(t)) = (A - LC)e(t) \end{aligned}$$

Therefore, if all the eigenvalues of $A - LC$ have negative real part, the estimation error $e(t)$ asymptotically tends to zero, and the state estimate $\hat{x}$ tends to x. Now, note that the eigenvalues of $A - LC$ are those of $A' - C'L'$ and the problem of assigning the eigenvalues of $A' - C'L'$ with a suitable choice of L' (or L) is equivalent to the one of computing the gain K to assign the eigenvalues of $A - BK$.

It is then possible to use the same algorithms discussed in the previous sections provided that the following substitutions are made

$$A \to A', \quad B \to C', \quad m \to p, \quad K \to L'$$

In particular, the reachablility condition

$$rank M_r = \begin{bmatrix} B & AB & \dots & A^{n-1}B \end{bmatrix} = n$$

becomes

$$rank M_o' = \begin{bmatrix} C' & A'C' & \dots & A^{n-1\prime}C' \end{bmatrix} = n$$

The matrix M_o is the observability matrix of the system, which must be observable.
The following result holds.

Theorem 6.2.1. *A necessary and sufficient condition for the design of an asymptotic observer with arbitrarily specified eigenvalues is that the pair (A, C) is observable.*

Remark. Assume that the system is just detectable, but not observable, i.e. the eigenvalues of the non observable part have negative real part. In this case, it is possible to build an observer whose eigenvalues are those of the nonobservable part, plus the remaining eigenvalues which can be placed in prescribed positions.

In the following it will always be assumed that the system is observable.

Systems with scalar output ($p = 1$)

Given the desired characteristic polynomial $P(s)$ associated with the observer's dynamics, the gain L can be computed with the Ackermann's formula which, in view of the previous considerations, takes the form

$$L' = \begin{bmatrix} 0 & 0 & \dots & 0 & 1 \end{bmatrix} M_o^{-1\prime} P(A')$$

or

$$L = P(A) M_o^{-1} \begin{bmatrix} 0 \\ 0 \\ \dots \\ 0 \\ 1 \end{bmatrix}$$

Systems with multiple outputs ($p > 1$)

Again, it is possible to use the same algorithms already introduced for the computation of the gain K which assigns the eigenvalues of $A - BK$ when $m > 1$, provided that the proper substitutions are made.

6.2.2 Observers for systems with disturbances

Assume that a disturbance d acts on the system's state and/or output, so that the system is described by

$$\begin{aligned} \dot{x}(t) &= Ax(t) + Bu(t) + Md(t) \\ y(t) &= Cx(t) + Du(t) + Nd(t) \end{aligned}$$

Measurable disturbance d

If d is measurable, it is possible to consider it as a known input signal, and the observer is

$$\dot{\hat{x}}(t) = A\hat{x}(t) + Bu(t) + Md(t) + L\left[y(t) - C\hat{x}(t) - Du(t) - Nd(t)\right]$$

With standard mathematical developments, analogous to those previously described, it is possible to see that the estimation error $e(t) = x(t) - \hat{x}(t)$ is the state of the following dynamical system

$$\dot{e}(t) = (A - LC)e(t)$$

All the previous considerations hold also in this case, so that $\hat{x}(t) \to x(t)$ when $t \to \infty$ provided that the eigenvalues of $(A - LC)$ have negative real part.

Nonmeasurable disturbance d

In this case, the observer cannot include the terms related to the disturbance, and is described by

$$\dot{\hat{x}}(t) = A\hat{x}(t) + Bu(t) + L\left[y(t) - C\hat{x}(t) - Du(t)\right]$$

It follows that

$$\dot{e}(t) = (A - LC)e(t) + (M - LN)d(t)$$

It is apparent that, due to the input d, the estimation error cannot tend to zero (unless d asymptotically vanishes) even if the observer is stable. For example, if d is constant, that is $d = \bar{d}$, the state estimation error tends to

$$\bar{e} = -(A - LC)^{-1}(M - LN)\bar{d}$$

Estimation of constant disturbances

Assume that the disturbance d is unknown, but constant, i.e. $d(t) = \bar{d}$. In this case, it is possible to estimate both the state x (without any bias) and the disturbance d. In fact, a simple procedure consists of modeling the disturbance as the state of the additional dynamic system

$$\dot{d}(t) = 0, \quad d(0) = \bar{d} \in R^r$$

The overall system is then described by

$$\begin{aligned} \begin{bmatrix} \dot{x}(t) \\ \dot{d}(t) \end{bmatrix} &= \begin{bmatrix} A & M \\ 0 & 0 \end{bmatrix} \begin{bmatrix} x(t) \\ d(t) \end{bmatrix} + \begin{bmatrix} B \\ 0 \end{bmatrix} u(t) \\ y(t) &= \begin{bmatrix} C & N \end{bmatrix} \begin{bmatrix} x(t) \\ d(t) \end{bmatrix} + Du(t) \end{aligned}$$

It is then possible to design an observer for this new system in order to estimate the enlarged state $[x'(t) \;\; d'(t)]'$ provided that the pair

$$\left(\begin{bmatrix} A & M \\ 0 & 0 \end{bmatrix}, \begin{bmatrix} C & N \end{bmatrix} \right)$$

is observable. It is easy to see that a necessary and sufficient condition for the observability of this pair is that (A, C) is observable and

$$rank \begin{bmatrix} A & M \\ C & N \end{bmatrix} = n + r$$

In turn, this means that the following relation is required to hold

$$r \leq p$$

so that the maximum number of disturbances which can be estimated is equal to the number of outputs.

6.2.3 Reduced order observers

Note that the observer previously introduced for the system

$$\begin{aligned} \dot{x}(t) &= Ax(t) + Bu(t) \\ y(t) &= Cx(t) + Du(t) \end{aligned}$$

is of order n. It is now shown how to design an observer of order $n - p$, where p is the number of outputs. For simplicity, in the following it will be assumed that $D = 0$ and the rank of C is p. It is then possible to find a (non unique) non singular and square matrix T with the following structure

$$T = \begin{bmatrix} C \\ T_1 \end{bmatrix}$$

where $T_1 \in R^{n-p,n}$ is a matrix which can be arbitrarily chosen provided that $\det(T) \neq 0$.
Now consider the following state transformation

$$\tilde{x} = Tx = \begin{bmatrix} C \\ T_1 \end{bmatrix} x \quad , \quad \tilde{x} = \begin{bmatrix} y \\ \tilde{x}_r \end{bmatrix}, \quad \tilde{x}_r \in R^{n-p}$$

and note that the first p components of the transformed state coincide with the system's outputs, which are known. Therefore, the problem reduces now to the estimation of the elements of the vector $\tilde{x}_r$ based on the measurements of the inputs u and of the outputs y.

By applying the previous state transformation to the original system, one obtains the following transformed system, in the new state variables $\tilde{x}$, described by the equations

$$\begin{aligned} \dot{\tilde{x}}(t) &= \tilde{A}\tilde{x}(t) + \tilde{B}u(t) \\ y(t) &= \tilde{C}x(t) \end{aligned}$$

where the matrices $\tilde{A} = TAT^{-1}$, $\tilde{B} = TB$, $\tilde{C} = CT^{-1}$ can be partitioned as follows

$$\tilde{A} = \begin{bmatrix} \tilde{A}_{11} & \tilde{A}_{12} \\ \tilde{A}_{21} & \tilde{A}_{22} \end{bmatrix}, \quad \tilde{A}_{11} \in R^{p,p}, \quad \tilde{A}_{22} \in R^{n-p,n-p}$$

$$\tilde{B} = \begin{bmatrix} \tilde{B}_1 \\ \tilde{B}_2 \end{bmatrix}, \quad \tilde{B}_1 \in R^{p,m}, \quad \tilde{B}_2 \in R^{n-p,m}$$

$$\tilde{C} = \begin{bmatrix} I_p & 0 \end{bmatrix}$$

Hence, the transformed system can be written in the form

$$\begin{aligned} \dot{y}(t) &= \tilde{A}_{11}y(t) + \tilde{A}_{12}\tilde{x}_r(t) + \tilde{B}_1u(t) \\ \dot{\tilde{x}}_r(t) &= \tilde{A}_{21}y(t) + \tilde{A}_{22}\tilde{x}_r(t) + \tilde{B}_2u(t) \end{aligned}$$

Now define the variables

$$\begin{aligned} \eta(t) &= \dot{y}(t) - \tilde{A}_{11}y(t) - \tilde{B}_1u(t) \\ \zeta(t) &= \tilde{A}_{21}y(t) + \tilde{B}_2u(t) \end{aligned}$$

and write this new system as

$$\begin{aligned} \dot{\tilde{x}}_r(t) &= \tilde{A}_{22}\tilde{x}_r(t) + \zeta(t) \\ \eta(t) &= \tilde{A}_{12}\tilde{x}_r(t) \end{aligned}$$

It is now possible to design an observer for the state $\tilde{x}_r$ by noting that the first equation is a standard state equation with known input $\zeta(t)$, made by the known or measurable terms u and y, while the second equation is a linear transformation expressing the dependence of the new output $\eta(t)$ on the state $\tilde{x}_r$. Moreover, it is possible to show that if the pair (A, C) is observable, then also the pair $(\tilde{A}_{22}, \tilde{A}_{12})$ is observable.

By means of the previous considerations, the estimate $\hat{x}_r$ of $\tilde{x}_r$ is given by

$$\dot{\hat{x}}_r(t) = \tilde{A}_{22}\hat{x}_r(t) + \zeta(t) + L\left[\eta(t) - \tilde{A}_{12}\hat{x}_r(t)\right]$$

or, in view of the definition of ζ and by means of a suitable reordering

$$\dot{\hat{x}}_r(t) = \left(\tilde{A}_{22} - L\tilde{A}_{12}\right)\hat{x}_r(t) + \tilde{A}_{21}y(t) + \tilde{B}_2u(t) + L\eta(t)$$

This expression shows that the gain matrix L can be chosen to assign the eigenvalues of $\left(\tilde{A}_{22} - L\tilde{A}_{12}\right)$, so as to guarantee the asymptotic stability of the observer.

However, the reduced order observer so obtained has input η, which in turn depends on the output derivative $\dot{y}$. This value can be numerically computed, but numerical errors or disturbances can prevent one from obtaining an acceptable accuracy. Alternatively, it is possible to further elaborate the previous observer's expression as follows. First, note that in view of the definition of η, an equivalent form of the observer is

$$\dot{\hat{x}}_r(t) - L\dot{y}(t) = (\tilde{A}_{22} - L\tilde{A}_{12})\hat{x}_r(t) + (\tilde{A}_{21} - L\tilde{A}_{11})y(t) + (\tilde{B}_2 - L\tilde{B}_1)u(t)$$

By summing and subtracting $(\tilde{A}_{22} - L\tilde{A}_{12})Ly(t)$ to the right hand side and by defining the variable

$$\xi(t) = \hat{x}_r(t) - Ly(t), \quad \xi \in R^{n-p}$$

one obtains

$$\dot{\xi}(t) = (\tilde{A}_{22} - L\tilde{A}_{12})\xi(t) + (\tilde{A}_{21} - L\tilde{A}_{11} + \tilde{A}_{22}L - L\tilde{A}_{12}L)y(t) + (\tilde{B}_2 - L\tilde{B}_1)u(t)$$

which is an easily implementable form of the reduced order observer where the term $\dot{y}$ does not appear anymore.

Once the value of $\xi(t)$ has been computed, the state estimate is

$$\hat{x}_r(t) = \xi(t) + Ly(t)$$

Moreover, it is possible to obtain the estimate $\hat{x}$ of the state x of the system in its original form by setting

$$\hat{x}(t) = T^{-1}\begin{bmatrix} y(t) \\ \hat{x}_r(t) \end{bmatrix}$$

6.3 Closed-loop system

By combining the model of the system

$$\begin{aligned} \dot{x}(t) &= Ax(t) + Bu(t) \\ y(t) &= Cx(t) + Du(t) \end{aligned}$$

with the asymptotic state observer

$$\dot{\hat{x}}(t) = A\hat{x}(t) + Bu(t) + L\left[y(t) - C\hat{x}(t) - Du(t)\right]$$

and with the control law applied to the state estimate

$$u(t) = -K\hat{x}(t) + v(t)$$

one obtains

$$\begin{aligned} \dot{x}(t) &= Ax(t) - BK\hat{x}(t) + Bv(t) \\ \dot{\hat{x}}(t) &= A\hat{x}(t) - BK\hat{x}(t) + Bv(t) + L\left[Cx(t) - C\hat{x}(t)\right] \end{aligned}$$

or, recalling that $e = x - \hat{x}$,

$$\begin{aligned} \dot{x}(t) &= (A - BK)x(t) + BKe(t) + Bv(t) \\ \dot{e}(t) &= (A - LC)e(t) \end{aligned}$$

This expression can be written in matrix form as follows

$$\begin{bmatrix} \dot{x}(t) \\ \dot{e}(t) \end{bmatrix} = \begin{bmatrix} A - BK & BK \\ 0 & A - LC \end{bmatrix} \begin{bmatrix} x(t) \\ e(t) \end{bmatrix} + \begin{bmatrix} B \\ 0 \end{bmatrix} v(t)$$

which shows that the eigenvalues of the closed-loop system are those of $A - BK$ and of $A - LC$. This proves the so-called *separation principle* which states that it is possible to independently design the static state-feedback control law (by assuming that the state measurement is available) and the observer, for example both with the pole assignment technique. By combining these two designs, the resulting overall closed-loop system has the eigenvalues previously independently fixed.

6.3.1 Regulator transfer function

The transfer function of the regulator can be computed by combining the observer

$$\dot{\hat{x}}(t) = A\hat{x}(t) + Bu(t) + L\left[y(t) - C\hat{x}(t) - Du(t)\right]$$

and the static control law

$$u(t) = -K\hat{x}(t) + v(t)$$

It follows that

$$\begin{aligned} \dot{\hat{x}}(t) &= (A - LC)\hat{x}(t) - (B - LD)K\hat{x}(t) + (B - LD)v(t) + Ly(t) \\ u(t) &= -K\hat{x}(t) + v(t) \end{aligned}$$

and, letting for simplicity,

$$\bar{A} = A - LC - BK + LDK, \quad \bar{B} = (B - LD)$$

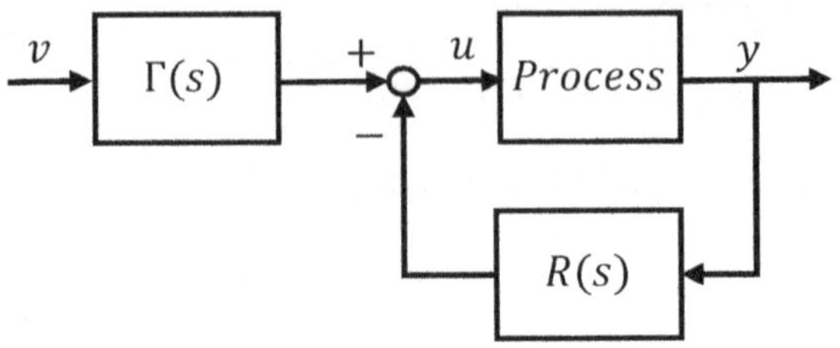

Figure 6.2: Control system with output feedback.

the regulator is

$$\begin{aligned}\dot{\hat{x}}(t) &= \bar{A}\hat{x}(t) + \bar{B}v(t) + Ly(t)\\ u(t) &= -K\hat{x}(t) + v(t)\end{aligned}$$

In Laplace transforms one has

$$U(s) = -K(sI - \bar{A})^{-1}LY(s) + \left[-K(sI - \bar{A})^{-1}\bar{B} + I\right] V(s)$$

The structure of the regulator is shown in Figure 6.2, where

$$R(s) = K(sI - \bar{A})^{-1}L, \quad \Gamma(s) = -K(sI - \bar{A})^{-1}\bar{B} + I$$

Remark. The dynamics of the regulator is due to the state observer, and its order is n, i.e. the order of the observer. In view of the results reported in Paragraph 6.2.3, this order can be reduced to $n - p$ if a reduced order observer is used.

6.4 Discrete-time systems

Given the system

$$\begin{aligned}x(k+1) &= Ax(k) + Bu(k), \quad x \in R^n, \quad u \in R^m\\ y(k) &= Cx(k) + Du(k), \quad y \in R^p\end{aligned}$$

the goal is to find a control law which places the eigenvalues of the closed-loop system in prescribed positions of the complex plane.

If the state is measurable, it is possible to use the static control law

$$u(k) = -Kx(k) + v(k), \quad K \in R^{m,n}, \quad v \in R^m$$

so that the closed-loop system is

$$x(k+1) = (A - BK)x(k) + Bv(k)$$

Then, the pole placement problem consists of finding a matrix K such that $A - BK$ has prescribed eigenvalues. This problem is the same already considered for continuous-time systems, and all the previous considerations hold (reachability assumption of the pair (A, B), Ackermann's formula, ...).

6.4.1 Nonmeasurable state

Different observers can be used to compute the estimate $\hat{x}$ of the true state x. In particular, the estimate of $x(k)$ can be based on:

- the measurements of the input u and of the output y up to time $k-1$; in this case, a *prediction problem* is solved, and a *state predictor* is used. The estimated state is denoted by $\hat{x}(k|k-1)$;
- the measurements of u and y up to time k; in this case a *filtering problem* is solved, and a *state filter* is used. The estimated state is denoted by $\tilde{x}(k|k)$.

State predictor

The predictor equation is

$$\hat{x}(k+1|k) = A\hat{x}(k|k-1) + Bu(k) + L\left[y(k) - C\hat{x}(k|k-1) - Du(k)\right]$$

and the dynamics of the estimation error $\hat{e}(k|k-1) = x(k) - \hat{x}(k|k-1)$ is given by

$$\hat{e}(k+1|k) = (A - LC)\hat{e}(k|k-1)$$

As in the continuous-time case, if the pair (A, C) is observable it is possible to compute a gain matrix L which places the eigenvalues of $A - LC$ in prescribed positions within the stability region. Note that, if the eigenvalues are placed at the origin of the complex plane, a *deadbeat observer* is used and the state estimation error goes to zero at most in n time steps.

State filter

The equation of the filter is

$$\begin{aligned} \tilde{x}(k+1|k+1) &= A\tilde{x}(k|k) + Bu(k) \\ &+ L\left[y(k+1) - C\left(A\tilde{x}(k|k) + Bu(k)\right) - Du(k+1)\right] \end{aligned}$$

Note that the ideal corrective term, which should be written as

$$L\left[y(k+1) - C\tilde{x}(k+1|k+1) - Du(k+1)\right],$$

is modified by using the best estimate of $x(k+1)$ which can be computed at time k. Letting $\tilde{e}(k|k) = x(k) - \tilde{x}(k|k)$, by means of simple computations one obtains

$$\tilde{e}(k+1|k+1) = (A - LCA)\tilde{e}(k|k)$$

It is then possible to arbitrarily assign the eigenvalues of the estimation error dynamics provided that the pair (A, CA) is observable. To this regard, note that the corresponding (transposed) observability matrix is

$$\begin{aligned} \tilde{M}_o' &= \left[\begin{array}{cccc} A'C' & A^{2\prime}C' & \dots & A^{n\prime}C' \end{array}\right] \\ &= A'\left[\begin{array}{cccc} C' & A'C' & \dots & A^{n-1\prime}C' \end{array}\right] = A'M_o' \end{aligned}$$

Hence, if the pair (A, C) is observable, the observability of (A, CA) may be lost only if A has null eigenvalues, otherwise the pair (A, CA) is only detectable and it is not possible to find a gain L which arbitrarily assigns all the observer's eigenvalues. However, the eigenvalues which cannot be moved are in $z = 0$, and correspond to the fastest possible dynamics.

6.4.2 Reduced order observer

A reduced order observer can be developed along the same lines adopted in the continuous-time case.

Given the system

$$\begin{aligned} x(k+1) &= Ax(k) + Bu(k) \\ y(k) &= Cx(k) \end{aligned}$$

where the pair (A, C) is observable and the rank of matrix C is p, define a transformation matrix T as follows

$$T = \begin{bmatrix} C \\ T_1 \end{bmatrix}$$

where $T_1 \in R^{n-p,n}$ is a matrix which can be arbitrarily chosen provided that $\det(T) \neq 0$.

By applying the state transformation

$$\tilde{x} = Tx = \begin{bmatrix} C \\ T_1 \end{bmatrix} x, \quad \tilde{x} = \begin{bmatrix} y \\ \tilde{x}_r \end{bmatrix}, \quad \tilde{x}_r \in R^{n-p}$$

the original system can be written as

$$\begin{aligned} \tilde{x}(k+1) &= \tilde{A}\tilde{x}(k) + \tilde{B}u(k) \\ y(k) &= \tilde{C}\tilde{x}(k) \end{aligned}$$

where $\tilde{A} = TAT^{-1}$, $\tilde{B} = TB$, $\tilde{C} = CT^{-1}$ can be partitioned as follows

$$\tilde{A} = \begin{bmatrix} \tilde{A}_{11} & \tilde{A}_{12} \\ \tilde{A}_{21} & \tilde{A}_{22} \end{bmatrix}, \quad \tilde{A}_{11} \in R^{p,p}, \quad \tilde{A}_{22} \in R^{n-p,n-p}$$

$$\tilde{B} = \begin{bmatrix} \tilde{B}_1 \\ \tilde{B}_2 \end{bmatrix}, \quad \tilde{B}_1 \in R^{p,m}, \quad \tilde{B}_2 \in R^{n-p,m}$$

$$\tilde{C} = \begin{bmatrix} I_p & 0 \end{bmatrix}$$

Correspondingly, the transformed system can be written as

$$\begin{aligned} y(k+1) &= \tilde{A}_{11}y(k) + \tilde{A}_{12}\tilde{x}_r(k) + \tilde{B}_1u(k) \\ \tilde{x}_r(k+1) &= \tilde{A}_{21}y(k) + \tilde{A}_{22}\tilde{x}_r(k) + \tilde{B}_2u(k) \end{aligned}$$

Define the variables

$$\begin{aligned}\eta(k+1) &= y(k+1) - \tilde{A}_{11}y(k) - \tilde{B}_1 u(k)\\ \zeta(k) &= \tilde{A}_{21}y(k) + \tilde{B}_2 u(k)\end{aligned}$$

and write the transformed system as

$$\begin{aligned}\tilde{x}_r(k+1) &= \tilde{A}_{22}\tilde{x}_r(k) + \zeta(k)\\ \eta(k+1) &= \tilde{A}_{12}\tilde{x}_r(k)\end{aligned}$$

It is now possible to build a state observer in order to compute the estimate of $\tilde{x}_r$. In fact, the observability of the pair (A, C) implies the observability of the pair $(\tilde{A}_{22}, \tilde{A}_{12})$ of the transformed and reduced system. Therefore, by means of the same considerations previously introduced, the estimate $\hat{x}_r$ of the state $\tilde{x}_r$ is given by

$$\hat{x}_r(k+1\,|k+1) = \tilde{A}_{22}\hat{x}_r(k\,|k) + \zeta(k) + L\left[\eta(k+1) - \tilde{A}_{12}\hat{x}_r(k\,|k)\right]$$

or, from the definition of ζ and η,

$$\begin{aligned}\hat{x}_r(k+1\,|k+1) &= \left(\tilde{A}_{22} - L\tilde{A}_{12}\right)\hat{x}_r(k\,|k) + \tilde{A}_{21}y(t) + \tilde{B}_2 u(t)\\ &+ L[y(k+1) - \tilde{A}_{11}y(k) - \tilde{B}_1 u(k)]\end{aligned}$$

This expression shows that the gain matrix L can be chosen to assign the eigenvalues of $\left(\tilde{A}_{22} - L\tilde{A}_{12}\right)$. Note that the observer has the structure of a *filter*, since the estimate of the state at a given time k depends on the output at the same time instant.

By computing at any time instant the value of $\hat{x}_r(k\,|k)$, it is then possible to determine the estimate $\hat{x}$ of x by setting

$$\hat{x}(k\,|k) = T^{-1}\begin{bmatrix} y(k)\\ \hat{x}_r(k\,|k)\end{bmatrix}$$

7
Optimal control

This chapter introduces a very general control synthesis technique for linear and nonlinear systems. The approach is based on the formulation, in the time domain, of the control problem in terms of an optimization one, where constraints on the state and input variables can also be considered.

In this chapter, the problem is formulated with reference to continuous-time, invariant and nonlinear systems, and sufficient solvability conditions are derived. In the following chapters, these results will be specialized to linear systems and the stability and robustness properties of the resulting control law will be derived. Furthermore, all the main results will be extended to discrete time systems.

7.1 Problem formulation

Consider the nonlinear, time invariant system

$$\dot{x}(t) = f(x(t), u(t)), \quad x \in R^n, \quad u \in R^m \tag{7.1}$$

where f is continuously differentiable with respect to its arguments, and the state x is measurable.

Assuming to be at the initial time instant t_0, we want to compute an "optimal control" $u^o(t)$, $t \in [t_0, T]$ minimizing the performance index (also called *cost function* in the following)

$$J(x(t_0), u(\cdot), t_0) = \int_{t_0}^{T} l(x(\tau), u(\tau))d\tau + m(x(T)) \tag{7.2}$$

where l and m are continuously differentiable functions.

The performance index (7.2) must be minimized subject to the constraints due to the system's dynamics (7.1) and to additional state and input constraints:

$$x(t) \in X, \quad u(t) \in U \tag{7.3}$$

where $X \subseteq R^n$ and $U \subseteq R^m$ are compact sets containing the origin.

Remark. In the previous formulation, a generic initial time t_0 has been considered; however, since we are dealing with time invariant systems, it is always possible to set $t_0 = 0$.

Remark. The design parameters are the prediction horizon T and the state and control weighting functions $l(x,u)$, $m(x)$. It is apparent that the problem solution will be *optimal* only with respect to these choices.

Remark. In order to show a typical choice of $l(x,u)$ and $m(x)$, consider a stabilization problem, where the goal is drive the state from $x(t_0)$ to the origin with a limited control energy. In this case, a reasonable choice is to set

$$l(x,u) = x'Qx + u'Ru \tag{7.4}$$
$$m(x) = x'Sx \tag{7.5}$$

where $Q \geq 0$ and $S \geq 0$ are (n,n) symmetric matrices and $R > 0$ is a symmetric matrix with dimensions (m,m). These choices weight the (squared) deviation of the state from the origin during the transient ($x'Qx$) and at the end of the prediction horizon ($x'Sx$), and penalize the control action ($u'Ru$). It is apparent that the relative weighting of these terms, i.e. the choice of Q, S, and R, allows one to express different requirements. As an example, consider the case of scalar state x and input u

$$\begin{aligned} x'Qx &\rightarrow qx^2 \\ u'Ru &\rightarrow ru^2 \\ x'Sx &\rightarrow sx^2 \end{aligned}$$

The selection of the scalars q, r, s must be made to describe the different goals of the control design. Performance indices given by (7.4), (7.5) are *quadratic* in x and u.

Example 7.1.1. Consider the RC circuit shown in Figure 7.1 and described by the model

$$C\dot{x}_c(t) = i(t)$$

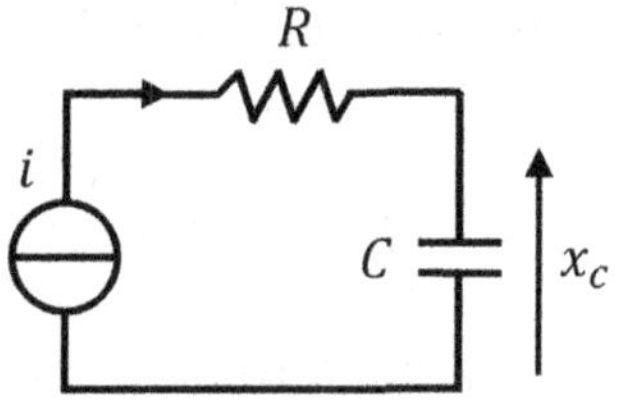

Figure 7.1: RC circuit

Given $x_c(t_0) = \bar{x}_c \neq 0$ and the prediction horizon T, the goal is to compute the current i in order to:

1. minimize the energy stored in the capacitor at time $t = T$ given a known initial condition;

2. minimize the power dissipated in the resistance.

A suitable performance index is

$$J = s x_c^2(T) + \int_{t_0}^{T} R i^2(\tau) d\tau, \quad s \geq 0$$

where s is a design parameter. It is apparent that the choice of s emphasizes the first (high s) or the second (small s) goal, and the optimality of the solution will be referred to the adopted value of the parameter s.

7.2 Problem solution

In the following, sufficient conditions will be derived under the assumption that the problem is unconstrained, i.e. that the state and control constraints (7.3) are missing.

Denote by $u_{[a,b]}$ the control functions $u(\cdot)$ in the interval $[a, b]$ and define

$$J^0(x(t), t) = \min_{u[t,T]} J(x(t), u(\cdot), t) = \int_t^T l(x(\tau), u(\tau)) d\tau + m(x(T))$$

where t is a generic time instant in the interval $[t_0, T]$.

Remark. Both J^0 and J depend on $x(t)$, i.e. on the current value of the state, while they do not depend on the state evolution up to time t.

Now recall the dynamic programming principle, also called Bellman's principle of Optimality:

From any point of an optimal trajectory, the remaining trajectory is optimal for the corresponding problem over the remaining number of stages, or time interval, initiated at that point.

Then, it is possible to write

$$J^0(x(t), t) = \min_{u[t,t_1]} \left\{ \min_{u[t_1,T]} \left[\int_t^{t_1} l(x(\tau), u(\tau)) d\tau + \int_{t_1}^{T} l(x(\tau), u(\tau)) d\tau + m(x(T)) \right] \right\}$$

In the previous expression, the term

$$\min_{u[t_1,T]} \left[\int_t^{t_1} l(x(\tau), u(\tau)) d\tau + \int_{t_1}^{T} l(x(\tau), u(\tau)) d\tau + m(x(T)) \right]$$

depends on the control applied in the interval $[t, t_1]$, and, as such, on the state at time t_1. However, the first term does not depend on $u[t_1, T]$, therefore

$$
\begin{aligned}
&J^0(x(t), t) \\
&= \min_{u[t,t_1]} \left\{ \int_t^{t_1} l(x(\tau), u(\tau))d\tau + \min_{u[t_1,T]} \left[\int_{t_1}^{T} l(x(\tau), u(\tau))d\tau + m(x(T)) \right] \right\}
\end{aligned}
$$

It follows that

$$
J^0(x(t), t) = \min_{u[t,t_1]} \left\{ \int_t^{t_1} l(x(\tau), u(\tau))d\tau + J^o(x(t_1), t_1) \right\}
$$

This expression shows that if the optimal control has been applied in the interval $[t_1, T)$, the optimal cost of the state trajectory starting at t is obtained by minimizing the sum of the cost incurred from t to t_1 plus the optimal cost from t_1 to T.

Now let $t_1 = t + dt$ in the previous expression, and note that, in view of the mean value theorem, there exists $\alpha \in [0, 1]$ such that

$$
J^0(x(t), t) = \min_{u[t,t+dt]} \{ l(x(t + \alpha dt), u(t + \alpha dt))dt + J^o(x(t + dt), t + dt) \}
$$

In view of the regularity assumptions previously introduced, it is possible to expand $J^o(x(t + dt), t + dt)$ as follows

$$
\begin{aligned}
&J^o(x(t + dt), t + dt) \\
&= J^o(x(t), t) + \frac{\partial J^o(x(t), t)}{\partial x} \frac{dx(t)}{dt} dt + \frac{\partial J^o(x(t), t)}{\partial t} dt + O(dt)^2
\end{aligned}
$$

Hence

$$
\begin{aligned}
&J^0(x(t), t) \\
&= \min_{u[t,t+dt]} \left\{ l(x(t + \alpha dt), u(t + \alpha dt))dt + J^o(x(t), t) + \frac{\partial J^o(x(t), t)}{\partial x} \frac{dx(t)}{dt} dt \right. \\
&\qquad \left. + \frac{\partial J^o(x(t), t)}{\partial t} dt + O(dt)^2 \right\}, \quad \alpha \in [0, 1]
\end{aligned}
$$

By simplifying, dividing by dt, and letting $dt \to 0$ one obtains

$$
0 = \min_{u[t]} \left\{ l(x(t), u(t)) + \frac{\partial J^o(x(t), t)}{\partial x} f(x(t), u(t)) + \frac{\partial J^o(x(t), t)}{\partial t} \right\}
$$

Note that $\dfrac{\partial J^o(x(t), t)}{\partial t}$ does not depend on u. Moreover, the previous expression must be evaluated at a given fixed time t, where x and u must be considered as vectors, instead of functions of time. As such, the previous expression can be written as

$$\frac{\partial J^o(x,t)}{\partial t} = -\min_u \left\{ l(x,u) + \frac{\partial J^o(x,t)}{\partial x} f(x,u) \right\} \tag{7.6}$$

where

$$J(x, u(\cdot), T) = m(x)$$

does not depend on the previous control, and in particular

$$J^o(x,T) = m(x) \tag{7.7}$$

Equations (7.6) and (7.7) are called the *Hamilton - Jacobi - Bellman* (HJB) equations of optimal control and have been used to formulate and solve many optimal control problems.

In view of the previous results, the solution of the optimization problem requires to proceed according to the following steps:

1. compute the value u^o minimizing

$$\left\{ l(x,u) + \frac{\partial J^o(x,t)}{\partial x} f(x,u) \right\}$$

   Assuming that there exists an unique minimum, this corresponds to compute a function κ with arguments x and $\dfrac{\partial J^o(x,t)}{\partial x}$

$$u^o = \kappa\left(x, \frac{\partial J^o(x,t)}{\partial x}\right)$$

2. compute the function $J^o(x,t)$ satisfying the HJB equation

$$\frac{\partial J^o(x,t)}{\partial t} = -l\left(x, \kappa\left(x, \frac{\partial J^o(x,t)}{\partial x}\right)\right) - \frac{\partial J^o(x,t)}{\partial x} f\left(x, \kappa\left(x, \frac{\partial J^o(x,t)}{\partial x}\right)\right)$$

   with the boundary condition

$$J^o(x,T) = m(x)$$

This second step is usually very difficult when nonlinear systems and general cost functions are considered, while in the following Chapter 8 a solution will be derived in the case of linear systems and quadratic performance indices.

Example 7.2.1. Consider the system

$$\dot{x}(t) = u(t)$$

and, letting $t_0 = 0$, the cost function

$$J(x(0), u(\cdot), 0) = \int_0^T \left(u(\tau)^2 + x(\tau)^2 + \frac{1}{2}x(\tau)^4 \right) d\tau$$

Since $l(x,u) = u^2 + x^2 + \frac{1}{2}x^4$ and $f(x,u) = u$, the HJB equation is

$$\frac{\partial J^o(x,t)}{\partial t} = -\min_u \left\{ u^2 + x^2 + \frac{1}{2}x^4 + \frac{\partial J^o(x,t)}{\partial x} u \right\}$$

By setting to zero the derivative of the cost function, it is easy to verify that the optimal control law is

$$u^o = \kappa\left(x, \frac{\partial J^o(x,t)}{\partial x}\right) = -\frac{1}{2}\frac{\partial J^o(x,t)}{\partial x}$$

The problem is now to compute a function $J^o(x,t)$ satisfying the HJB equation

$$\begin{aligned}\frac{\partial J^o(x,t)}{\partial t} &= -\frac{1}{4}\left(\frac{\partial J^o(x,t)}{\partial x}\right)^2 - x^2 - \frac{1}{2}x^4 + \frac{1}{2}\left(\frac{\partial J^o(x,t)}{\partial x}\right)^2 \\ &= \frac{1}{4}\left(\frac{\partial J^o(x,t)}{\partial x}\right)^2 - x^2 - \frac{1}{2}x^4\end{aligned}$$

under the condition

$$J^o(x,T) = 0$$

Remark. Note that, although the optimization problem has been defined as an open-loop one with respect to the variable u, if it is possible to compute the solution J^o of the HJB equation, the resulting control law $u^o = \kappa(x,t)$ is a closed-loop one. The corresponding scheme is shown in Figure 7.2.

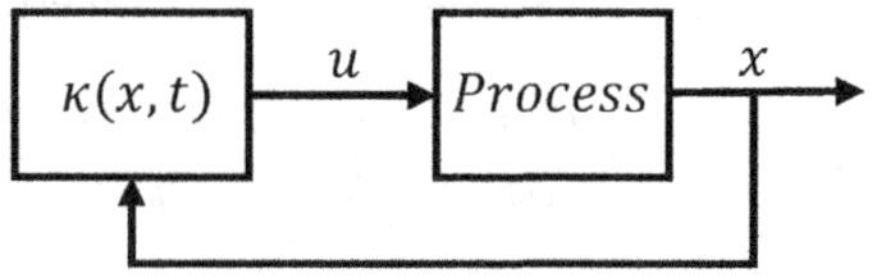

Figure 7.2: Closed-loop scheme with the optimal control law.

Remark. The optimal value of the cost function is $J^o(x(t_0), t_0)$.

Remark. The regularity of the functions f, l, and the uniqueness of the minimum of $l(x,u) + \frac{\partial J^o(x,t)}{\partial x} f(x,u)$ are sufficient conditions for the solution of the optimal control problem.

Remark. The previous developments can be easily extended to the case of time varying functions f and l.

8

Linear Quadratic control

In this chapter, the optimal control problem is formulated and solved for linear systems and quadratic cost functions, leading to the popular *Linear Quadratic*, or *LQ*, control approach. The properties of the *LQ* control law are studied in terms of stability and robustness of the resulting closed-loop system.

8.1 Finite horizon *LQ* control

Consider the following linear time invariant system, with initial time $t_0 = 0$ and measurable state

$$\dot{x}(t) = Ax(t) + Bu(t), \quad x(0) = x_0$$

and the performance index

$$J(x_0, u(\cdot), 0) = \int_0^T \left(x'(\tau)Qx(\tau) + u'(\tau)Ru(\tau)\right) d\tau + x'(T)Sx(T)$$

where

$$Q = Q' \geq 0, \quad S = S' \geq 0, \quad R = R' > 0$$

are design parameters, together with the terminal time instant T.

Remark. The assumption that Q, R, and S are symmetric is not fundamental, since given a non symmetric matrix Γ, it holds that

$$\begin{aligned} x'\Gamma x &= x'\frac{\Gamma + \Gamma'}{2}x + x'\frac{\Gamma - \Gamma'}{2}x \\ &= x'\frac{\Gamma + \Gamma'}{2}x + x'\frac{\Gamma}{2}x - x'\frac{\Gamma'}{2}x \end{aligned}$$

The last term is a scalar, and is equal to its transpose, so that the previous expression can be written as a symmetric one

$$\begin{aligned} x'\Gamma x &= x'\frac{\Gamma + \Gamma'}{2}x + x'\frac{\Gamma}{2}x - x'\frac{\Gamma}{2}x \\ &= x'\frac{\Gamma + \Gamma'}{2}x \end{aligned}$$

Therefore, it is equivalent to use a nonsymmetric matrix Γ or its symmetric part $\frac{\Gamma+\Gamma'}{2}$.

Remark. The assumption $R > 0$ implies that all the components of the control vector are weighted in J, so as to avoid the trivial solution consisting in the use of an impulsive control action driving the state to the origin in an arbitrary small time interval. Moreover, as it will be shown, this condition is also required to derive an explicit algorithm for the solution of the optimization problem.

In view of the HJB equation, the solution of the LQ problem is computed according to the following steps.

1. Compute the control u solving the problem

$$\min_u \left\{ l(x,u) + \frac{\partial J^o(x,t)}{\partial x} f(x,u) \right\}$$

In the considered case one has

$$\min_u \left\{ x'Qx + u'Ru + \frac{\partial J^o(x,t)}{\partial x} (Ax + Bu) \right\}$$

Since the term to be minimized is quadratic in u and positive definite due to the assumption $R > 0$, the minimum is obtained by setting to zero the partial derivative with respect to u, i.e.

$$2u'R + \frac{\partial J^o(x,t)}{\partial x} B = 0$$

or

$$u^o = -\frac{1}{2} R^{-1} B' \left(\frac{\partial J^o(x,t)}{\partial x} \right)'$$

2. Compute the function $J^o(x,t)$ fulfilling the HJB equation

$$\begin{aligned} \frac{\partial J^o(x,t)}{\partial t} &= -x'Qx - \frac{1}{4} \left(\frac{\partial J^o(x,t)}{\partial x} \right) BR^{-1}B' \left(\frac{\partial J^o(x,t)}{\partial x} \right)' \\ &\quad - \left(\frac{\partial J^o(x,t)}{\partial x} \right) \left[Ax - \frac{1}{2} BR^{-1}B' \left(\frac{\partial J^o(x,t)}{\partial x} \right)' \right] \end{aligned}$$

or

$$\begin{aligned} -\frac{\partial J^o(x,t)}{\partial t} &= x'Qx - \frac{1}{4} \left(\frac{\partial J^o(x,t)}{\partial x} \right) BR^{-1}B' \left(\frac{\partial J^o(x,t)}{\partial x} \right)' \\ &\quad + \left(\frac{\partial J^o(x,t)}{\partial x} \right) Ax \end{aligned} \qquad (8.1)$$

with the boundary condition

$$J^o(x,T) = x'Sx$$

A tentative expression of $J^o(x,t)$ is

$$J^o(x,t) = x'P(t)x$$

where $P(t)$ is a symmetric matrix such that

$$P(T) = S$$

This choice satisfies the boundary condition of the HJB equation

$$J^o(x,T) = x'Sx = m(x)$$

Moreover

$$\begin{aligned}\frac{\partial J^0(x,t)}{\partial t} &= \frac{\partial x'P(t)x}{\partial t} = x'\frac{dP(t)}{dt}x = x'\dot{P}(t)x \\ \frac{\partial J^0(x,t)}{\partial x} &= \frac{\partial x'P(t)x}{\partial x} = 2x'P(t)\end{aligned}$$

Hence, eq. (8.1) is equivalent to

$$\begin{aligned}-x'\dot{P}(t)x &= x'Qx - x'P(t)BR^{-1}B'P'(t)x + 2x'P(t)Ax \\ &= x'Qx - x'P(t)BR^{-1}B'P'(t)x + 2x'\frac{P(t)A + A'P'(t)}{2}x\end{aligned}$$

or

$$\dot{P}(t) + Q - P(t)BR^{-1}B'P'(t) + P(t)A + A'P'(t) = 0$$

this equation is known as the *differential Riccati equation.*

The corresponding control law is

$$u^o(t) = -R^{-1}B'P'(t)x(t) = -K(t)x(t)$$

where

$$K(t) = R^{-1}B'P'(t)$$

Remark. The assumption $R > 0$ guarantees that the control law is well defined.

Remark. In view of the symmetry of the Riccati equation and of the initialization ($P(T) = S = S'$), the matrix $P(t)$ is symmetric for any t, and in the following the symbol of transposition will be omitted. Moreover, since it has been set $J^0(x,t) = x'P(t)x$ and J^0 is always greater or equal to zero (sum of nonnegative quantities, since $Q \geq 0$, $S \geq 0$, and $R > 0$), $P(t)$ is nonnegative definite.

Remark. The computed control law is time-varying and defined in the interval $[0,T]$. However, the computation of the gain $K(t)$ does not require the knowledge of the current state, and can be made a-priori.

Remark. From the definition of $K(t)$, straightforward computations allow one to verify that the Riccati equation can be written in the equivalent form

$$\dot{P}(t) + Q + P(t)\left(A - BK(t)\right) + \left(A' - K'(t)B'\right)P(t) + K'(t)RK(t) = 0$$

8.2 Infinite horizon *LQ* control

From a practical point of view, a time-varying control law defined only over a finite interval is often not useful. In order to compute a time-invariant control law it is natural to define an *infinite horizon optimal control problem* by considering $T \to \infty$. In so doing, provided that the state x is properly weighted in the cost function J, and in order to guarantee that J is finite, it must turn out that $x \to 0$, i.e. the resulting control system must be asymptotically stable.

In view of these considerations, the infinite horizon *LQ* control problem consists of minimizing with respect to u the performance index

$$\begin{aligned} J(x_0, u(\cdot), 0) &= \int_0^{\infty} (x'(\tau)Qx(\tau) + u'(\tau)Ru(\tau))\, d\tau \\ & Q = Q' \geq 0, \quad R = R' > 0 \end{aligned}$$

The problem can be derived from the finite horizon one by setting

$$T \to \infty, \quad S = 0$$

Note that the condition $S = 0$ can be easily interpreted: if the resulting control system is asymptotically stable, the state must tend to zero as the time tends to infinity and the terminal penalty of the finite horizon formulation vanishes.

For the infinite horizon *LQ* problem the following result holds:

Theorem 8.2.1. *If the pair (A, B) is reachable:*

1. *the solution of the differential Riccati equation with initial condition $P(T) = 0$ tends, for $T \to \infty$, to a constant matrix $\bar{P} \geq 0$ solution of the algebraic Riccati equation*
$$0 = A'\bar{P} + \bar{P}A + Q - \bar{P}BR^{-1}B'\bar{P}$$
2. *the asymptotic control law is*
$$u(t) = -R^{-1}B'\bar{P}x(t) = -\bar{K}x(t) \tag{8.2}$$
where $\bar{K} = R^{-1}B'\bar{P}$.

Proof. Consider the finite horizon control problem

$$J^o(x(t), t, T) = \min_{u[t,T]} \int_t^T (x(\tau)'Qx(\tau) + u(\tau)'Ru(\tau))\, d\tau$$

where T is not assumed to be fixed.

As already shown, the solution of this problem can be computed through the Riccati equation

$$\begin{aligned} &-\dot{P}(t,T) = A'P(t,T) + P(t,T)A + Q - P(t,T)BR^{-1}B'P(t,T) \\ &P(T,T) = 0 \end{aligned}$$

Correspondingly, the optimal cost J, for any generic t and $\bar{T} > t$ is given by

$$J^o(x(t), t, \bar{T}) = x'(t)P(t, \bar{T})x(t) = \min_{u[t,\bar{T}]} \int_t^{\bar{T}} (x(\tau)'Qx(\tau) + u(\tau)'Ru(\tau))\, d\tau$$

Since the system is assumed to be reachable, there exist a finite $\bar{T} > t$ and a control $\bar{u}$ defined in $[t, \bar{T}]$ which drives the state to the origin $\forall t > \bar{T}$. Letting

$$u(\tau) = \begin{cases} \bar{u}(\tau), & \tau \in [t, \bar{T}] \\ 0, & \tau > \bar{T} \end{cases}$$

the cost function is

$$\bar{J}(x(t), t, T) = \int_t^{\bar{T}} (x'(\tau)Qx(\tau) + \bar{u}'(\tau)R\bar{u}(\tau))\, d\tau$$

In fact, from $\bar{T}$ to T both the state x and the control u are null. The value of $\bar{J}(x(t), t, T)$ is an upper bound for $J^o(x(t), t, T)$. Moreover $J^o(x(t), t, T)$ is monotonic and non decreasing function of T. In fact, if it is not, given $\bar{T}$, there should be $\tilde{T} > \bar{T}$ such that $J^o(x(t), t, \tilde{T}) < J^o(x(t), t, \bar{T})$. Now assume to apply the optimal input for $\tilde{T}$ up to $\bar{T}$: since the integrand is nonnegative, the value of J^o up to $\bar{T}$ is less than the one up to $\tilde{T}$, which contradicts the hypothesis. Therefore, since $J^o(x(t), t, T)$ is upper bounded and nondecreasing with respect to T, it must have limit for $T \to \infty$, i.e. $P(t, T) \to \bar{P}(t)$, $T \to \infty$. Recall that $J^o(x(t), t, T) = x'(t)P(t, T)x(t)$.

Matrix $\bar{P}(t)$ solves the Riccati equation, and in view of the continuity of this equation and the assumptions $Q \geq 0$, $R > 0$, it is symmetric and positive semidefinite.

In order to show that $\bar{P}(t)$ is constant, note that, for a given initial state x_0 in view of the stationarity of the system, and of the matrices Q, R, the optimal value of the performance index is the same for the problem defined over $[t_1, \infty]$ or over $[t_2, \infty]$, with $t_2 \neq t_1$. Hence

$$x_0'\bar{P}(t_1)x_0 = x_0'\bar{P}(t_2)x_0, \quad \forall x_0$$

and

$$\bar{P}(t_1) = \bar{P}(t_2) = \bar{P}$$

Remark. The computed control law is time invariant.

Remark. The reachability assumption of the pair (A, B) can be relaxed to the weaker assumption of stabilizability.

Although the choice $T \to \infty$ allows one to obtain the time invariant control law (8.2), still it is not guaranteed that the closed-loop system

$$\dot{x}(t) = Ax(t) - B\bar{K}x(t) = \left(A - B\bar{K}\right)x(t)$$

is asymptotically stable.

As already noted, in order to show that the control law (8.2) stabilizes the system it is necessary to guarantee that all the components of the state vector are weighted in the cost function J. In fact, in this case J^o cannot be bounded if one or more elements of the state vector do not asymptotically tend to zero, i.e. if the closed-loop system is not asymptotically stable. In view of these considerations, it is apparent that the choice of the matrix Q is fundamental to guarantee stability. A more formal justification of these qualitative considerations is now given.

First note that the matrix Q can always be partitioned as

$$Q = C_q' C_q$$

This partition is not unique: for example, if $Q = 1$ possible partitions are $C_q = \pm 1$.

Now observe that, assuming for simplicity $t_0 = 0$ and $\tilde{y}(\tau) = C_q x(\tau)$, it is possible to write

$$\begin{aligned} J(x_0, u(\cdot), 0) &= \int_0^\infty \left(x'(\tau) C_q' C_q x(\tau) + u'(\tau) R u(\tau) \right) d\tau \\ &= \int_0^\infty \left(\tilde{y}'(\tau) \tilde{y}(\tau) + u'(\tau) R u(\tau) \right) d\tau \end{aligned}$$

According to this interpretation, the performance index penalizes a “fictitious” output $\tilde{y}$. It is then apparent that the requirement to penalize the state can be formulated as an observability property when the fictitious output $\tilde{y}$ is considered, that is in terms of observability of the pair (A, C_q).

In order to derive the main stability result, the following intermediate Lemmas must be introduced.

Lemma 8.2.2. *Let C_{q1} and C_{q2} be two partitions of Q, i.e.*

$$Q = C_{q1}' C_{q1} = C_{q2}' C_{q2} \tag{8.3}$$

Then, if (A, C_{q1}) is observable, also (A, C_{q2}) is observable.

Proof. Since by assumption (A, C_{q1}) is observable,

$$rank(M_{01}) = rank \left(\begin{bmatrix} C_{q1} \\ C_{q1} A \\ \vdots \\ C_{q1} A^{n-1} \end{bmatrix} \right) = n$$

Hence, there does not exist $x_0 \neq 0$ such that $M_{01} x_0 = 0$. Assume now that (A, C_{q2}) is not observable, then there exists $x_0 \neq 0$ such that $M_{02} x_0 = 0$, that is

$$\begin{bmatrix} C_{q2} \\ C_{q2} A \\ \vdots \\ C_{q2} A^{n-1} \end{bmatrix} x_0 = 0$$

This is equivalent to

$$\begin{aligned}
C_{q2}x_0 &= 0 \rightarrow x_0'C_{q2}'C_{q2}x_0 = 0 \\
C_{q2}Ax_0 &= 0 \rightarrow x_0'A'C_{q2}'C_{q2}Ax_0 = 0 \\
&\vdots \\
C_{q2}A^{n-1}x_0 &= 0 \rightarrow x_0'A^{n-1\prime}C_{q2}'C_{q2}A^{n-1}x_0 = 0
\end{aligned}$$

Recalling (8.3), these relations are equivalent to

$$\begin{aligned}
x_0'C_{q1}'C_{q1}x_0 &= 0 \\
x_0'A'C_{q1}'C_{q1}Ax_0 &= 0 \\
&\vdots \\
x_0'A^{n-1\prime}C_{q1}'C_{q1}A^{n-1}x_0 &= 0
\end{aligned}$$

that is

$$M_{01}x_0 = 0, \quad x_0 \neq 0$$

which contradicts the initial assumption.

In view of the previous result, when considering the observability property it will be possible to use any generic partition C_q of the matrix Q.

Lemma 8.2.3. *The asymptotic solution $\bar{P}$ of the Riccati equation is positive definite if and only if (A, C_q) is observable.*

Proof. **Sufficiency** Assume that (A, C_q) is observable and, by contradiction, $\bar{P} \geq 0$. Then, there exists an initial state $x_0 \neq 0$ such that

$$J^0(x_0) = x_0'\bar{P}x_0 = 0$$

Since $R > 0$, in order to have $J^0 = 0$ it must hold that $u(t) = 0$, $\forall t \geq 0$. Correspondingly, the controlled system is

$$\dot{x}(t) = Ax(t)$$

and the state evolves according to

$$x(t) = e^{At}x_0$$

Correspondingly, the optimal cost is

$$\begin{aligned}
J^o(x_0) &= \int_0^\infty x'(\tau)Qx(\tau)d\tau \\
&= x_0' \int_0^\infty e^{A\tau'}C_q'C_qe^{A\tau}d\tau x_0 = x_0'W_ox_0 = 0
\end{aligned}$$

where

$$W_o = \int_0^\infty e^{A\tau'}C_q'C_qe^{A\tau}d\tau$$

is the so-called observability gramian. This conditions contradicts the observability test based on the observability gramian (recall that a necessary and sufficient condition for the observability of the pair (A, C_q) is $W_0 > 0$).

Necessity Let $\bar{P} > 0$. If the pair (A, C_q) is not observable, setting $u = 0$ there should exist $x_0 \neq 0$ such that $C_q e^{At} x_0 = 0$. Hence

$$J^o(x_0) = \int_0^\infty x_0' e^{A\tau'} C_q' C_q e^{A\tau} x_0 d\tau = 0$$

and $u = 0$ would be the optimal control. In this case

$$x_0' \bar{P} x_0 = 0$$

would imply $\bar{P} \geq 0$ since $x_0 \neq 0$.

Remark. If $Q > 0$, C_q has full rank n and the observability of the pair (A, C_q) is trivially verified. Since Q is a project parameter, this consideration shows that it is easy to verify the observability assumption by properly choosing the design parameter Q.

It is now possible to state the fundamental stability result of *LQ* control.

Theorem 8.2.4. *If*

1. *the pair (A, B) is reachable;*
2. *the pair (A, C_q) is observable*

then

A) *the optimal control law is given by*

$$u(t) = -\bar{K}x(t)$$

with

$$\bar{K} = R^{-1}B'\bar{P}$$

where $\bar{P}$ is the unique positive definite solution of the stationary Riccati equation

$$0 = \bar{P}A + A'\bar{P} + Q - \bar{P}BR^{-1}B'\bar{P}$$

B) *the closed-loop system*

$$\dot{x}(t) = (A - B\bar{K})x(t)$$

is asymptotically stable.

Proof. Considering item A), it is only necessary to prove the uniqueness of the solution $\bar{P} > 0$ of the Riccati equation; this problem will be studied once item B) has been proved.

Stability It is proven that

$$J^o(x) = x'\bar{P}x$$

is a Lyapunov function. In fact, $J^o(x) > 0, \forall x \neq 0$ since, in view of Lemma 8.2.3, $\bar{P} > 0$. Moreover

$$\begin{aligned}
\frac{\partial J^o(x)}{\partial t} &= \dot{x}'\bar{P}x + x'\bar{P}\dot{x} \\
&= x'(A - B\bar{K})'\bar{P}x + x'\bar{P}(A - B\bar{K})x \\
&= x'\left\{A'\bar{P} - \bar{K}'B'\bar{P} + \bar{P}A - \bar{P}B\bar{K}\right\}x \\
&= x'\left\{A'\bar{P} - \bar{P}BR^{-1}B'\bar{P} + \bar{P}A - \bar{P}BR^{-1}B'\bar{P}\right\}x \\
&= -x'\left\{Q + \bar{P}BR^{-1}B'\bar{P}\right\}x \\
&= -x'\left\{Q + \bar{K}'R\bar{K}\right\}x
\end{aligned}$$

If $Q > 0$ the proof is completed in view of the Lyapunov theorem. If $Q \geq 0$ it is necessary to show that $\frac{\partial J^o(x)}{\partial t}$ is not identically null along the trajectory $\bar{x}(t)$ with initial condition $\bar{x}_0 \neq 0$ in order to apply the Krasowski-La Salle lemma (Theorem 1.3.3). Assume by contradiction that $\frac{\partial J^o(\bar{x}(t))}{\partial t} = 0$, that is

$$-\bar{x}'(t)\left\{Q + \bar{P}BR^{-1}B'\bar{P}\right\}\bar{x}(t) = 0, \quad \forall t$$

or

$$\bar{x}'(t)Q\bar{x}(t) = 0, \quad \forall t$$

and

$$\bar{x}'(t)\bar{P}BR^{-1}B'\bar{P}\bar{x}(t) = 0, \quad \forall t$$

From this relation one can conclude that

$$B'\bar{P}\bar{x}(t) = 0, \quad \forall t$$

Hence

$$u(t) = -R^{-1}B'\bar{P}\bar{x}(t) = 0, \quad \forall t$$

and the state of the closed-loop system evolves according to its free motion

$$\bar{x}(t) = e^{At}\bar{x}_0, \quad \bar{x}_0 \neq 0$$

Therefore

$$\bar{x}'(t)Q\bar{x}(t) = \bar{x}_0'e^{A't}C_q'C_qe^{At}\bar{x}_0 = 0, \quad \forall t, \quad \bar{x}_0 \neq 0$$

which is not possible in view of the observability assumption of the pair (A, C_q). In fact, recall that the observability gramian

$$W_o = \int_0^\infty \left(e^{A'\tau}C_q'C_qe^{A\tau}\right)d\tau$$

of an observable system is positive definite.

Consider now item (A), and assume that the stationary Riccati equation has two positive definite solutions $\bar{P}_1 > 0$ and $\bar{P}_2 > 0$. In view of the previous proof,

$A_{cl1} = (A - B\bar{K}_1)$ and $A_{cl2} = (A - B\bar{K}_2)$, where $\bar{K}_1 = R^{-1}B'\bar{P}_1$, $\bar{K}_2 = R^{-1}B'\bar{P}_2$, have eigenvalues with negative real part. Note that

$$\begin{aligned} 0 &= \bar{P}_1 A + A'\bar{P}_1 + Q - \bar{P}_1 B R^{-1} B'\bar{P}_1 \\ 0 &= \bar{P}_2 A + A'\bar{P}_2 + Q - \bar{P}_2 B R^{-1} B'\bar{P}_2 \end{aligned}$$

It then follows that

$$\begin{aligned} & (\bar{P}_1 - \bar{P}_2)A_{cl1} + A'_{cl2}(\bar{P}_1 - \bar{P}_2) \\ = \; & (\bar{P}_1 - \bar{P}_2)\left(A - BR^{-1}B'\bar{P}_1\right) + \left(A - BR^{-1}B'\bar{P}_2\right)'(\bar{P}_1 - \bar{P}_2) \\ = \; & \bar{P}_1 A - \bar{P}_1 BR^{-1}B'\bar{P}_1 - \bar{P}_2 A + \bar{P}_2 BR^{-1}B'\bar{P}_1 + A'\bar{P}_1 - A'\bar{P}_2 \\ & -\bar{P}_2 BR^{-1}B'\bar{P}_1 + \bar{P}_2 BR^{-1}B'\bar{P}_2 \\ = \; & \left(\bar{P}_1 A + A'\bar{P}_1 - \bar{P}_1 BR^{-1}B'\bar{P}_1\right) - \left(\bar{P}_2 A + A'\bar{P}_2 - \bar{P}_2 BR^{-1}B'\bar{P}_2\right) \\ = \; & Q - Q = 0 \end{aligned}$$

Hence

$$(\bar{P}_1 - \bar{P}_2)A_{cl1} + A'_{cl2}(\bar{P}_1 - \bar{P}_2) = 0$$

However, the matrix equation

$$FX + XG = 0$$

has an unique solution $X = 0$ whenever

$$\lambda_i(F) + \lambda_j(G) \neq 0, \quad \forall i, j$$

Since, with reference to A_{cl1} and A_{cl2} this condition is fulfilled, it follows that $\bar{P}_1 - \bar{P}_2 = 0$, this leads to conclude that the solution is unique.

Remark. The solution of the stationary Riccati equation can be computed by integration or (better) with suitable numerical algorithms.

Example 8.2.1. Consider the first order unstable system

$$\dot{x}(t) = x(t) + u(t)$$

where $A = 1$, $B = 1$.

The stationary Riccati equation is

$$2\bar{P} + Q - \frac{\bar{P}^2}{R} = 0$$

and

$$\bar{P} = \frac{2 \pm \sqrt{4\left(1 + \frac{Q}{R}\right)}}{2/R}$$

Now consider the following choices.

1. $Q = 3$, $R = 1$ $\left(\frac{Q}{R} = 3\right)$

$$\begin{aligned}\bar{P} &= \frac{2 \pm \sqrt{16}}{2} = \begin{cases} 3 & solution > 0 \\ -1 & \end{cases} \\ &\bar{K} = R^{-1}B'\bar{P} = 3 \\ &A - B\bar{K} = 1 - 3 = -2\end{aligned}$$

2. $Q = 8$, $R = 1$ $\left(\frac{Q}{R} = 8\right)$

$$\begin{aligned}\bar{P} &= \frac{2 \pm \sqrt{36}}{2} = \begin{cases} 4 & solution > 0 \\ -2 & \end{cases} \\ &\bar{K} = R^{-1}B'\bar{P} = 4 \\ &A - B\bar{K} = 1 - 4 = -3\end{aligned}$$

3. $Q = 15$, $R = 1$ $\left(\frac{Q}{R} = 15\right)$

$$\begin{aligned}\bar{P} &= \frac{2 \pm \sqrt{64}}{2} = \begin{cases} 5 & solution > 0 \\ -3 & \end{cases} \\ &\bar{K} = R^{-1}B'\bar{P} = 5 \\ &A - B\bar{K} = 1 - 5 = -4\end{aligned}$$

Therefore, as also shown in Figure 8.1, when the ratio Q/R increases, also the speed of response of the feedback system with *LQ* control increases.

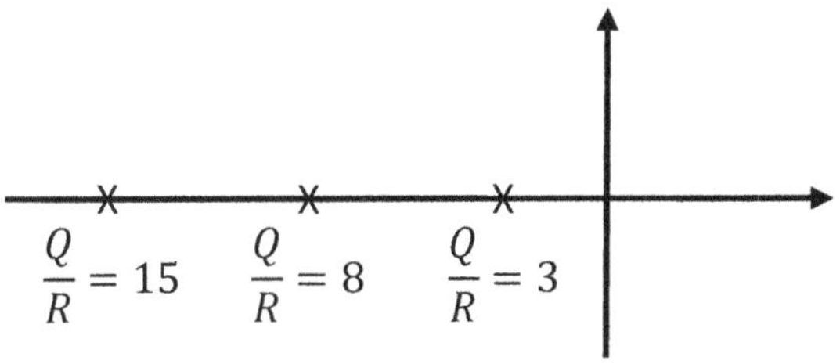

Figure 8.1: Eigenvalues of the feedback system of Example 8.2.1 corresponding to different choices of Q/R.

Example 8.2.2. Consider the first order system with two inputs and described by $A = a$, $B = [b_1\ b_2]$

$$\dot{x}(t) = ax(t) + b_1 u_1(t) + b_2 u_2(t)$$

Solve the infinite horizon *LQ* problem with the following weighting matrices

$$Q = q, \quad R^{-1} = \begin{bmatrix} \rho_1 & 0 \\ 0 & \rho_2 \end{bmatrix}$$

The algebraic Riccati equation is

$$0 = 2\bar{p}a + q - \bar{p}^2\left(b_1^2\rho_1 + b_2^2\rho_2\right)$$

If, for example, we set $a = 0$ (the system is an integrator) and $b_1 = b_2 = 1$, it follows that

$$\begin{aligned}
\bar{p} &= \sqrt{\frac{q}{\rho_1+\rho_2}} \\
\bar{K} &= R^{-1}B'\bar{p} = \begin{bmatrix} \rho_1 & 0 \\ 0 & \rho_2 \end{bmatrix}\begin{bmatrix} 1 \\ 1 \end{bmatrix}\sqrt{\frac{q}{\rho_1+\rho_2}} \\
&= \begin{bmatrix} \rho_1 \\ \rho_2 \end{bmatrix}\sqrt{\frac{q}{\rho_1+\rho_2}} \\
&= \begin{bmatrix} k_1 \\ k_2 \end{bmatrix} = \begin{bmatrix} \sqrt{\frac{\rho_1^2 q}{\rho_1+\rho_2}} \\ \sqrt{\frac{\rho_2^2 q}{\rho_1+\rho_2}} \end{bmatrix}
\end{aligned}$$

If $\rho_1 \gg \rho_2$, that is the weight of u_1 is smaller than the one of u_2 (recall that ρ_1 and ρ_2 are the elements of R^{-1}), the gain k_1 associated with u_1 turns out to be greater than the gain k_2 associated with u_2 (and viceversa).

8.3 Choice of the design parameters

As it has been shown in the previous example, the choice of the design parameters Q and R has a strong influence on the solution. Since this choice can be difficult in nontrivial cases, in the following two methods are described to select Q and R according to simple rules.

8.3.1 The normalization approach

Given

$$Q = \begin{bmatrix} q_1 & 0 & 0 \\ 0 & \ddots & 0 \\ 0 & 0 & q_n \end{bmatrix}, \quad q_i \geq 0, \quad R = \begin{bmatrix} r_1 & 0 & 0 \\ 0 & \ddots & 0 \\ 0 & 0 & r_m \end{bmatrix}, \quad r_i > 0$$

one has

$$J = \int_0^\infty \left(q_1 x_1^2(\tau) + \ldots + q_n x_n^2(\tau) + r_1 u_1^2(\tau) + \ldots + r_m u_m^2(\tau)\right) d\tau$$

Assuming to know that $|x_i| < x_{\max_i}$, $i = 1, \ldots, n$ and $|u_i| < u_{\max_i}$, $i = 1, \ldots, m$, and letting

$$q_i = \frac{\tilde{q}_i}{x_{\max_i}^2}, i = 1, \ldots, n, \quad r_i = \frac{\tilde{r}_i}{u_{\max_i}^2}, i = 1, \ldots, m, \quad \tilde{q}_i \geq 0, \quad \tilde{r}_i > 0$$

the performance index is

$$J = \int_0^\infty \left(\tilde{q}_1 \frac{x_1^2(\tau)}{x_{\max_1}^2} + \ldots + \tilde{q}_n \frac{x_n^2(\tau)}{x_{\max_n}^2} + \tilde{r}_1 \frac{u_1^2(\tau)}{u_{\max_1}^2} + \ldots + \tilde{r}_m \frac{u_m^2(\tau)}{u_{\max_m}^2} \right) d\tau$$

where $\tilde{q}_i$ and $\tilde{r}_i$ weight quantities belonging to the interval (0,1), so that their choice is independent on the adopted measurement units.

8.3.2 Weighting the outputs

Assume that the system has a measurable state and an output transformation

$$\begin{aligned} \dot{x}(t) &= Ax(t) + Bu(t) \\ y(t) &= Cx(t) \end{aligned}$$

with $y \in R^p$. If the design goal is to weight the outputs, a reasonable choice is

$$Q = C'\bar{Q}C, \quad \bar{Q} \geq 0, \quad \bar{Q} \in R^{p,p}$$

In fact

$$\begin{aligned} J &= \int_0^\infty (x'(\tau)Qx(\tau) + u'(\tau)Ru(\tau))\, d\tau \\ &= \int_0^\infty \left(x'(\tau)C'\bar{Q}Cx(\tau) + u'(\tau)Ru(\tau)\right) d\tau \\ &= \int_0^\infty (y'(\tau)\bar{Q}y(\tau) + u'(\tau)Ru(\tau))\, d\tau \end{aligned}$$

In view of the choice of $\bar{Q}$ and R the previous considerations hold.

8.3.3 *LQ* control with prescribed rate of convergence

It is possible to modify the ***LQ*** design to guarantee that the eigenvalues of $(A - B\bar{K})$ have negative real part smaller than $-\alpha$, where α is an arbitrary positive scalar.

Consider the following performance index

$$J(x_0, u(\cdot), 0) = \int_0^\infty e^{2\alpha\tau} (x'(\tau)Qx(\tau) + u'(\tau)Ru(\tau))\, d\tau$$

In order to solve the problem, define

$$\begin{aligned} \tilde{x}(t) &= e^{\alpha t}x(t) \\ \tilde{u}(t) &= e^{\alpha t}u(t) \end{aligned}$$

so that

$$\begin{aligned}\dot{\tilde{x}}(t) &= \alpha e^{\alpha t}x(t) + e^{\alpha t}\dot{x}(t) = \alpha\tilde{x}(t) + e^{\alpha t}\left(Ax(t) + Bu(t)\right) \\ &= \alpha\tilde{x}(t) + A\tilde{x}(t) + B\tilde{u}(t) = \left(\alpha I + A\right)\tilde{x}(t) + B\tilde{u}(t) \\ &= \tilde{A}\tilde{x}(t) + B\tilde{u}(t)\end{aligned}$$

where $\tilde{A} = \alpha I + A$. Moreover, the cost function can be written as

$$J(x_0, u(\cdot), 0) = \int_0^\infty \left(\tilde{x}'(\tau)Q\tilde{x}(\tau) + \tilde{u}'(\tau)R\tilde{u}(\tau)\right) d\tau$$

Therefore, the *LQ* problem is transformed into a standard one for the system with state $\tilde{x}$ and input $\tilde{u}$. Note that if (A, B) is reachable, $(\tilde{A}, B)$ is reachable as well. Then, the solution of the stated problem is given by

$$\tilde{u}(t) = -R^{-1}B'\bar{P}_\alpha\tilde{x}(t) = -\bar{K}_\alpha\tilde{x}(t)$$

where, if (A, B) is reachable and (A, C_q) is observable, $\bar{P}_\alpha$ is the unique positive definite solution of

$$0 = \bar{P}_\alpha\tilde{A} + \tilde{A}'\bar{P}_\alpha - \bar{P}_\alpha BR^{-1}B'\bar{P}_\alpha + Q$$

and all the eigenvalues of

$$\tilde{A} - B\bar{K}_\alpha = A + \alpha I - B\bar{K}_\alpha$$

have negative real part. This means that the eigenvalues of $A - B\bar{K}_\alpha$ have real part less then $-\alpha$.

In conclusion, the algorithm evolves according to the following steps.

1. Build the matrix $\tilde{A} = A + \alpha I$.

2. Compute the matrix $\bar{K}_\alpha$, solution of the *LQ* problem, for the system described by the matrices $\tilde{A}$ and B.

3. Implement the control law

$$u(t) = -\bar{K}_\alpha x(t)$$

Example 8.3.1. Given the unstable system

$$\dot{x}(t) = x(t) + u(t)$$

the goal is to obtain a closed-loop system with an eigenvalue smaller than $-\alpha$.

The system matrices are $A = B = 1$. Letting

$$\tilde{A} = A + \alpha = 1 + \alpha$$

the Riccati equation is

$$0 = 2\bar{P}\tilde{A} + Q - \bar{P}^2R^{-1}$$

or

$$\frac{\bar{P}^2}{R} - 2\,(1+\alpha)\,\bar{P} - Q = 0$$

with

$$\bar{P} = \frac{2\,(1+\alpha) + \sqrt{4\,(1+\alpha)^2 + 4\frac{Q}{R}}}{2/R} = \frac{(1+\alpha) + \sqrt{(1+\alpha)^2 + \frac{Q}{R}}}{1/R}$$

Therefore

$$\begin{aligned} \bar{K}_\alpha &= R^{-1}B\bar{P} = (1+\alpha) + \sqrt{(1+\alpha)^2 + \frac{Q}{R}} \\ A - B\bar{K}_\alpha &= -\alpha - \sqrt{(1+\alpha)^2 + \frac{Q}{R}} \end{aligned}$$

and, as required, the eigenvalue is smaller than $-\alpha$.

8.4 Robustness of *LQ* control

In this section, some results concerning the robustness properties of infinite horizon *LQ* control will be derived. Specifically, gain and phase variations of the controlled systems will be considered. For simplicity, the analysis will be completed only for single input systems, while only the main results will be reported for systems with multiple inputs ($m > 1$). Moreover, in order to simplify the notation, the matrices $\bar{P}$ and $\bar{K}$ will be denoted by P and K, respectively, and it will be assumed that all the hypotheses guaranteeing the stability of the closed-loop system with *LQ* control are fulfilled.

Recall that the optimal control law is

$$u = -Kx$$

where $K = R^{-1}B'P$, and P is the unique positive definite solution of the algebraic Riccati equation

$$A'P + PA + Q - PBR^{-1}B'P = 0$$

or

$$Q = -A'P - PA + K'RK$$

By adding and subtracting sP to the right hand side one obtains

$$Q = (-sI - A')P + P(sI - A) + K'RK$$

Now multiply this expression from the left by $B'(-sI - A')^{-1}$ and from the right by $(sI - A)^{-1}B$

$$\begin{aligned} B'(-sI - A')^{-1}Q(sI - A)^{-1}B &= B'P(sI - A)^{-1}B + B'(-sI - A')^{-1}PB \\ &\quad + B'(-sI - A')^{-1}K'RK(sI - A)^{-1}B \end{aligned}$$

From the definition of K one obtains

$$RK = B'P$$

and

$$K'R = PB$$

Hence, the previous expression can be written as

$$\begin{aligned} B'(-sI - A')^{-1}Q(sI - A)^{-1}B &= RK(sI - A)^{-1}B + B'(-sI - A')^{-1}K'R \\ &\quad + B'(-sI - A')^{-1}K'RK(sI - A)^{-1}B \end{aligned}$$

By adding and subtracting R from the two members, it is easy to show that the previous expression can be written as

$$\begin{aligned} & R + B'(-sI - A')^{-1}Q(sI - A)^{-1}B \\ = & \left\{I + B'(-sI - A')^{-1}K'\right\} R \left\{I + K(sI - A)^{-1}B\right\} \end{aligned}$$

Define

$$\begin{aligned} G_c(s) &= (sI - A)^{-1}B \\ \Gamma(s) &= I + K(sI - A)^{-1}B \end{aligned}$$

Then, the computed expression can be given the form

$$G_c'(-s)QG_c(s) + R = \Gamma'(-s)R\Gamma(s)$$

Therefore

$$\Gamma^*(j\omega)R\Gamma(j\omega) \geq R$$

where $\Gamma^*(j\omega) = \Gamma'(-j\omega)$.
Now note that the system and the control law can be described by the scheme of Figure 8.2.

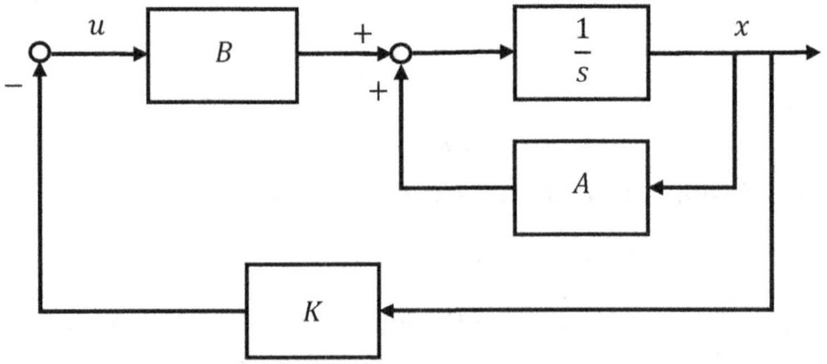

Figure 8.2: Feedback system.

It turns out that $G_c(s)$ is the transfer matrix from u to x, while, denoting by $L(s) = K(sI - A)^{-1}B$ the loop transfer function, $\Gamma(s) = (I + L(s))$ is the inverse of the sensitivity function (or *return difference*) computed by breaking the loop at the input u.

Case $m = 1$
In the scalar case $L(j\omega)$ is a transfer function, R is a scalar and it holds that

$$(1 + L(j\omega))^* R(1 + L(j\omega)) \geq R$$

that is

$$|1 + L(j\omega)|^2 \geq 1$$

Hence, the polar diagram of $L(j\omega)$ is always external to the circle centered in -1 and with radius one, as shown in Figure 8.3.

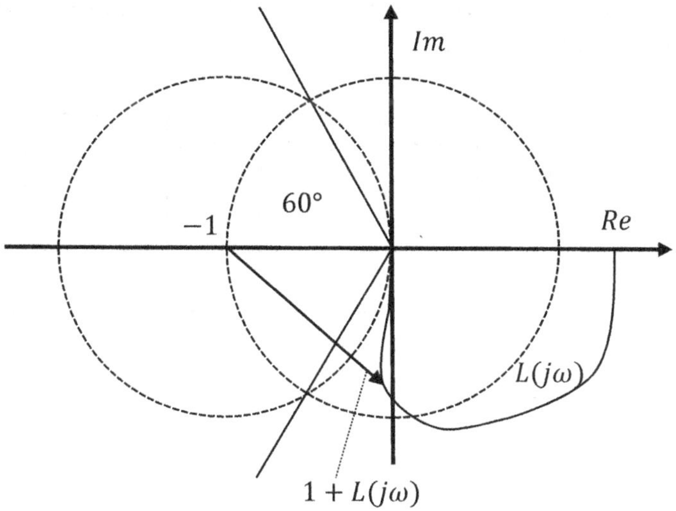

Figure 8.3: Robustness of LQ control.

In view of these considerations, the intersections between the Nyquist diagram and the circle centered at the origin and with radius one can only be characterized by angles in the interval $(-\frac{2}{3}\pi, \frac{2}{3}\pi)$ (unless multiples of 2π). Moreover, intersections with the negative real axis can only be to the left of -2 or with nonnegative abscissas. In view of the stability of the feedback system, due to the stabilizing property of *LQ* control, this means that the feedback system is asymptotically stable against phase variations of $\mp 60^0$ or gain variations in the interval $(1/2, \infty)$. For these reasons, the *LQ* control guarantees a phase margin at least equal to 60^0 and a gain margin $(1/2, \infty)$.

Note however that these results *do not mean* that the feedback system is robustly stable for *contemporary* phase and gain variations.

As for the high frequency behavior of the feedback system, note that, by recalling the relationship between the sensitivity ($S(s)$) and the complementary sensitivity ($T(s)$) functions, by means of the matrix inversion Lemma (see Appendix A), it is possible to show that

$$\begin{aligned} T(j\omega) &= 1 - S(j\omega) \\ &= 1 - \left[1 + K(j\omega I - A)^{-1}B\right]^{-1} \\ &= K(j\omega I - A + BK)^{-1}B \end{aligned}$$

Hence

$$\lim_{\omega \to \infty} j\omega T(j\omega) = KB = -R^{-1}B'PB$$

and $|T(j\omega)|$ decreases at high frequency with slope -1. This is not a nice behavior, since it would be better to have a loop transfer function $L(j\omega)$ with a lower slope at high frequency in order to reduce the effects of measurement noise and/or high frequency unmodelled dynamics.

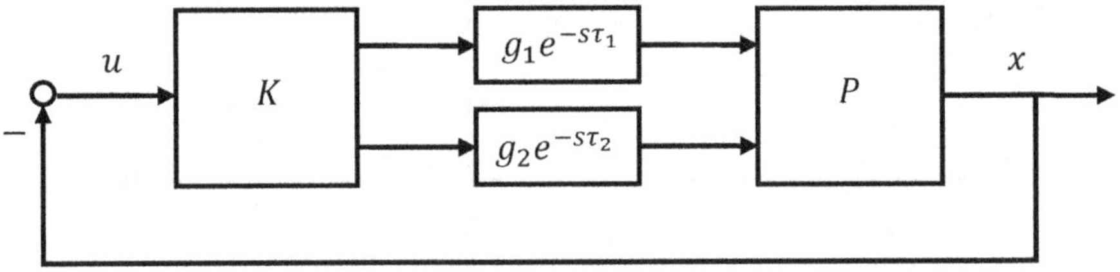

Figure 8.4: Gain and phase perturbations allowed by *LQ* control.

Case $m > 1$

In the case of multiple input systems, the previous results can be generalized to conclude that, if $R = diag\{r_1, r_2, \ldots, r_m\}$, $r_i > 0$, with reference to *each input* the closed-loop system remains stable in front of

1. phase variations of magnitude up to 60^0;
2. gain variations $(1/2, \infty)$.

However, as for $m = 1$, there are no guarantees for contemporary gain and phase variations in any input. This is schematically represented, for $m = 2$, in Figure 8.4, where g_i represents the gain variation (nominal value $g_i = 1$) and the delay $e^{-s\tau_i}$ represents the phase variation (nominal value $\tau_i = 0$).

9
Kalman filter

This chapter presents the Kalman Filter (*KF*), a state observer with optimality properties for systems subject to stochastic disturbances. The main stability and robustness properties of the *KF* are derived by exploiting its duality with respect to the *LQ* control described in the previous chapter.

9.1 Time varying Kalman filter

Consider the system

$$\begin{aligned} \dot{x}(t) &= Ax(t) + Bu(t) + v_x(t) \\ y(t) &= Cx(t) + v_y(t) \end{aligned}$$

where $x \in R^n$, $y \in R^p$, $u \in R^m$ and $v = \begin{bmatrix} v_x \\ v_y \end{bmatrix}$ is a white gaussian noise with expected value equal to zero and covariance matrix V, that is

$$E\left[v(t)\right] = 0, \quad E\left[v(t_1)v(t_2)'\right] = V\delta(t_1 - t_2)$$

where δ is the Kroenecker index and

$$V = \begin{bmatrix} \tilde{Q} & Z \\ Z' & \tilde{R} \end{bmatrix}, \quad \tilde{Q} \in R^{n,n}, \quad \tilde{R} \in R^{p,p}, \quad Z \in R^{n,p}$$

In the following, it will be always assumed that $\tilde{Q} \geq 0$, $\tilde{R} > 0$, i.e. there is a noise component with non zero variance acting on each output. Moreover, for simplicity it will be initially assumed that the noises acting on the state and on the output are uncorrelated, that is $Z = 0$.

The initial state $x_0 = x(0)$ is modelled as a Gaussian random variable with expected value $\bar{x}_0$ and covariance matrix

$$E\left[(x_0 - \bar{x}_0)(x_0 - \bar{x}_0)'\right] = \tilde{P}_0 \geq 0$$

It is also assumed that x_0 and the noises v_x, v_y are uncorrelated, that is

$$E\left[x_0 \begin{bmatrix} v_x' & v_y' \end{bmatrix}\right] = 0$$

For this system, consider now the following observer

$$\dot{\hat{x}}(t) = A\hat{x}(t) + Bu(t) + L(t)\left[y(t) - C\hat{x}(t)\right]$$

where $\hat{x} \in R^n$ and $L(t) \in R^{n,p}$ is a time varying gain to be chosen according to an optimality criterion.
From the previous equations, the dynamics of the error $e(t) = x(t) - \hat{x}(t)$ is

$$\begin{aligned}\dot{e}(t) &= Ae(t) - L(t)\left[Ce(t) + v_y(t)\right] + v_x(t) \\ &= \left[A - L(t)C\right]e(t) + B_c(t)v(t) \\ &= A_c(t)e(t) + B_c(t)v(t)\end{aligned}$$

where

$$A_c(t) = \left[A - L(t)C\right], \quad B_c(t) = \begin{bmatrix} I & -L(t) \end{bmatrix}$$

By applying the expectation operator and setting $\bar{e}(t) = E\left[e(t)\right]$, since $E\left[v(t)\right] = 0$, it follows that

$$\dot{\bar{e}}(t) = A_c(t)\bar{e}(t)$$

Therefore, the expected value of the estimation error is the state of a linear autonomous system and, letting $\hat{x}(0) = \bar{x}_0$, one has

$$E\left[e(0)\right] = E\left[x(0) - \hat{x}(0)\right] = 0$$

and

$$\bar{e}(t) = 0, \forall t \geq 0$$

In the following, this assumption will always be made, i.e. the initial state of the observer will be chosen so that the expected value of the error is always null.

Now define by $\tilde{P}(t)$ the covariance matrix of the estimation error; since $\bar{e}(t) = 0$,

$$\tilde{P}(t) = E\left[e(t)e(t)'\right]$$

with $\tilde{P}(0) = \tilde{P}_0$.

The goal is now to choose the observer gain $L(t)$ to minimize a quadratic form which depends on the covariance matrix of the error, that is to solve the following problem

$$\min_{L(t)} \gamma'\tilde{P}(t)\gamma$$

where $\gamma \in R^{n,1}$ is a generic vector. The following result holds.

Theorem 9.1.1. *The solution of the above problem is*

$$L(t) = \tilde{P}(t)C'\tilde{R}^{-1}$$

where $\tilde{P}(t)$ is the solution of the Riccati equation

$$\dot{\tilde{P}}(t) = A\tilde{P}(t) + \tilde{P}(t)A' + \tilde{Q} - \tilde{P}(t)C'\tilde{R}^{-1}C\tilde{P}(t)$$

with initial condition

$$\tilde{P}(0) = \tilde{P}_0$$

Proof. Since

$$\dot{e}(t) = A_c(t)e(t) + B_c(t)v(t)$$

one has

$$e(t) = \phi(t,0)e(0) + \int_0^t \phi(t,\tau)B_c(\tau)v(\tau)d\tau$$

where $\phi(t,\tau)$ is the transition matrix fulfilling the following equation

$$\frac{d\phi(t,\tau)}{dt} = A_c(t)\phi(t,\tau), \quad \phi(0,0) = I$$

Then

$$\begin{aligned} & E\left[e(t_1)e(t_2)'\right] \\ = \; & \phi(t_1,0)E\left[e(0)e(0)'\right]\phi(t_2,0)' && (9.1) \\ & +E\left[\phi(t_1,0)e(0)\left(\int_0^{t_2}\phi(t_2,\tilde{\tau})B_c(\tilde{\tau})v(\tilde{\tau})d\tilde{\tau}\right)'\right] && (9.2) \\ & +E\left[\int_0^{t_1}\phi(t_1,\tau)B_c(\tau)v(\tau)d\tau\,(\phi(t_2,0)e(0))'\right] && (9.3) \\ & +E\left[\int_0^{t_1}\phi(t_1,\tau)B_c(\tau)v(\tau)d\tau\left(\int_0^{t_2}\phi(t_2,\tilde{\tau})B_c(\tilde{\tau})v(\tilde{\tau})d\tilde{\tau}\right)'\right] && (9.4) \end{aligned}$$

In this expression, the terms (9.2) and (9.3) are null, since $e(0)$ and $v(t)$ are uncorrelated.

The term (9.4) can be written as

$$\begin{aligned} & E\left[\int_0^{t_1} d\tau \left(\int_0^{t_2}\phi(t_1,\tau)B_c(\tau)v(\tau)v(\tilde{\tau})'B_c(\tilde{\tau})'\phi(t_2,\tilde{\tau})d\tilde{\tau}\right)\right] \\ = \; & \int_0^{t_1} d\tau \left(\int_0^{t_2}\phi(t_1,\tau)B_c(\tau)E\left[v(\tau)v(\tilde{\tau})'\right]B_c(\tilde{\tau})'\phi(t_2,\tilde{\tau})d\tilde{\tau}\right) \\ = \; & \int_0^{t_1} d\tau \left(\int_0^{t_2}\phi(t_1,\tau)B_c(\tau)VB_c(\tilde{\tau})'\phi(t_2,\tilde{\tau})\delta(\tau-\tilde{\tau})d\tilde{\tau}\right) \\ = \; & \int_0^{\tilde{t}}\phi(t_1,\tau)B_c(\tau)VB_c(\tau)'\phi(t_2,\tau)d\tau \end{aligned}$$

where $\tilde{t} = \min(t_1,t_2)$.

Hence, reformulating also the term (9.1) and letting $t_1 = t_2 = t$, one has

$$\begin{aligned} \tilde{P}(t) &= E\left[e(t)e(t)'\right] = \phi(t,0)\tilde{P}(0)\phi(t,0)' \\ &\quad + \int_0^t \phi(t,\tau)B_c(\tau)VB_c(\tau)'\phi(t,\tau)'d\tau \end{aligned}$$

Differentiating with respect to time this expression, it follows that

$$\begin{aligned} \frac{d\left(\phi(t,0)\tilde{P}(0)\phi(t,0)'\right)}{dt} &= A_c(t)\phi(t,0)\tilde{P}(0)\phi(t,0)' \\ &\quad +\phi(t,0)\tilde{P}(0)\phi(t,0)'A_c'(t) \end{aligned}$$

and

$$\begin{aligned}
&\frac{d}{dt}\left[\int_0^t \phi(t,\tau)B_c(\tau)VB_c(\tau)'\phi(t,\tau)'d\tau\right] \\
&= \int_0^t A_c(t)\phi(t,\tau)B_c(\tau)VB_c(\tau)'\phi(t,\tau)'d\tau \\
&+ \int_0^t \phi(t,\tau)B_c(\tau)VB_c(\tau)'\phi(t,\tau)'A_c(t)'d\tau \\
&+ B_c(t)VB_c(t)'
\end{aligned}$$

Therefore

$$\dot{\tilde{P}}(t) = A_c(t)\tilde{P}(t) + \tilde{P}(t)A_c'(t) + B_c(t)VB_c(t)'$$

Recalling the definition of A_c and B_c, one has

$$\begin{aligned}
A_c(t)\tilde{P}(t) &= (A - L(t)C)\,\tilde{P}(t) \\
B_c(t)VB_c(t)' &= \begin{bmatrix} I & -L(t) \end{bmatrix}\begin{bmatrix} \tilde{Q} & 0 \\ 0 & \tilde{R} \end{bmatrix}\begin{bmatrix} I \\ -L(t)' \end{bmatrix} = \tilde{Q} + L(t)\tilde{R}L(t)'
\end{aligned}$$

The update of the error covariance is then given by

$$\begin{aligned}
\dot{\tilde{P}}(t) &= (A - L(t)C)\,\tilde{P}(t) + \tilde{P}(t)\,(A - L(t)C)' + \tilde{Q} + L(t)\tilde{R}L(t)' \\
\tilde{P}(0) &= \tilde{P}_0
\end{aligned}$$

Now note that, given the system

$$\dot{x}(t) = f(t, x(t)), \quad t \geq t_0, \quad x(t_0) = x_0$$

and the system

$$-\dot{y}(t) = f(t^* - t, y(t)), \quad t \leq t_1, \quad y(t_1) = y_1$$

with $t_1 > t_0$ and $t^* = t_0 + t_1$, if $x_0 = y_1$ the solutions of the two problems fulfill the equations

$$\begin{aligned}
x(t) &= y(t^* - t), \quad t \geq t_0 \\
y(t) &= x(t^* - t), \quad t \leq t_1
\end{aligned}$$

In order to prove this result, one can simply set $t \to t^* - t$.

Now consider the previous Riccati equation, with reversed time, and re-write it in terms of the variable $\hat{P}$

$$\begin{aligned}
-\dot{\hat{P}}(t) &= (A - L(t^* - t)C)\,\hat{P}(t) + \hat{P}(t)\,(A - L(t^* - t)C)' + \tilde{Q} \\
&\quad + L(t^* - t)\tilde{R}L(t^* - t)'
\end{aligned}$$

where $t^* = t_0 + t_1 = 0 + t_1$, $t_1 > 0$, and with the boundary condition $\hat{P}(t_1) = \tilde{P}_0$. From the previous result, one has

$$\tilde{P}(t) = \hat{P}(t^* - t), \quad t \leq t_1$$

Now, considering a time ranging from t_1 to $t_0 = 0$ and letting

$$\begin{array}{ccc} A & \longleftrightarrow & A' \\ Q & \longleftrightarrow & \tilde{Q} \\ R & \longleftrightarrow & \tilde{R} \\ B & \longleftrightarrow & C' \\ K(t) & \longleftrightarrow & L'(t^* - t) \\ P(t) & \longleftrightarrow & \hat{P}(t) \end{array}$$

one obtains the Riccati equation (8.1), in the variable $\hat{P}$, solving the LQ control problem written as

$$\min_{K(t)} x'P(t)x$$

whose solution is

$$K(t) = R^{-1}B'P(t)$$

Turning back to the problem of minimizing the functional $\gamma'\tilde{P}(t)\gamma$ with respect to $L(t)$, and in view of the previous considerations, one has

$$L(t^* - t)' = \tilde{R}^{-1}C\hat{P}(t) = \tilde{R}^{-1}C\tilde{P}(t^* - t)$$

If the time is reversed again, the result follows.

The proof of the previous theorem clearly shows that the properties of the observer can be directly derived from those of LQ control by means of duality considerations and with the following change of variables

LQ	KF
A	A'
B	C'
Q	$\tilde{Q}$
R	$\tilde{R}$
P	$\tilde{P}$
t	$t^* - t$
K	L'

9.2 Steady state Kalman filter

These considerations can be extended to obtain asymptotic results similar to those derived for infinite horizon LQ control. In particular, it is possible to compute a time invariant observer gain with stabilizing properties. The following result is equivalent to the one reported in the previous Chapter, and, for this reason, it is not proven here.

Theorem 9.2.1. *If*

1. *the pair* (A, B_q) *is reachable, where* B_q *is such that* $\tilde{Q} = B_q B_q'$;
2. *the pair* (A, C) *is observable;*

   then

A) *the optimal estimator is*

$$\begin{aligned} \dot{\hat{x}}(t) &= A\hat{x}(t) + Bu(t) + \bar{L}\left[y(t) - C\hat{x}(t)\right] \\ &= \left(A - \bar{L}C\right)\hat{x}(t) + Bu(t) + \bar{L}y(t) \end{aligned}$$

with

$$\bar{L} = \overline{\tilde{P}}C'\tilde{R}^{-1}$$

where $\overline{\tilde{P}}$ *is the unique positive definite solution of the stationary Riccati equation*

$$0 = A\overline{\tilde{P}} + \overline{\tilde{P}}A' + \tilde{Q} - \overline{\tilde{P}}C'\tilde{R}^{-1}C\overline{\tilde{P}}$$

B) *the observer is asymptotically stable, that is all the eigenvalues of* $\left(A - \bar{L}C\right)$ *have negative real part.*

Remark. When the noises are correlated, i.e. $Z \neq 0$, it is possible to write the state equation as

$$\begin{aligned} \dot{x}(t) &= Ax(t) + Bu(t) + v_x(t) + Z\tilde{R}^{-1}(y(t) - y(t)) \\ &= Ax(t) + Bu(t) + v_x(t) + Z\tilde{R}^{-1}y(t) - Z\tilde{R}^{-1}Cx(t) \\ &\quad - Z\tilde{R}^{-1}v_y(t) \\ &= \left(A - Z\tilde{R}^{-1}C\right)x(t) + Bu(t) + Z\tilde{R}^{-1}y(t) + \bar{v}(t) \end{aligned}$$

where $\bar{v}(t) = v_x(t) - Z\tilde{R}^{-1}v_y(t)$. It follows that

$$\begin{aligned} E\left[\bar{v}(t)v_y(t)'\right] &= Z - Z\tilde{R}^{-1}\tilde{R} = 0 \\ E\left[\bar{v}(t)\bar{v}(t)'\right] &= \tilde{Q} - Z\tilde{R}^{-1}Z' = \bar{\tilde{Q}} \geq 0 \end{aligned}$$

since the covariance matrix V is positive semidefinite by definition.

Then, letting

$$\bar{A} = A - Z\tilde{R}^{-1}C$$

the system

$$\begin{aligned} \dot{x}(t) &= \bar{A}x(t) + Bu(t) + Z\tilde{R}^{-1}y(t) + \bar{v}(t) \\ y(t) &= Cx(t) + v_y(t) \end{aligned}$$

fulfills the initial assumption of uncorrelated noises $\bar{v}$ and v_y. Note that the term $Z\tilde{R}^{-1}y(t)$ does not influence the estimation procedure, since it can be handled as any other known input.

Remark. Restricting the analysis to the considered linear structure of the observer does not limit the generality of the result. In fact, it is possible to prove that the observer previously developed is optimal, due to the initial assumptions of gaussianity.

Remark. It is often difficult to know the exact value of the covariances $\tilde{Q}$ and $\tilde{R}$. In particular, $\tilde{R}$ can be obtained by recalling that its diagonal elements are the variances of the available output measurements, so that they depend on the quality of the sensors. On the contrary, the estimate of $\tilde{Q}$ is usually more difficult. From another point of view, $\tilde{Q}$ and $\tilde{R}$ can be considered as free design parameters, similar to the matrices Q and R in LQ control. According to this interpretation, the performance of the observer, i.e. the speed of convergence of the state estimate to the true value, depends on the "ratio" between $\tilde{Q}$ and $\tilde{R}$: a matrix $\tilde{R}$ "bigger" than $\tilde{Q}$, corresponds to very noisy output measurements, and the filter will be "slow". Viceversa, $\tilde{R}$ "smaller" than $\tilde{Q}$ corresponds to the fact that the noise acting on the state is larger than the one on the output, and the filter will be "fast".

Example 9.2.1. Consider the problem of estimating a constant $\bar{x}$ with only noisy measurements available. It is possible to define a "fictitious" dynamics for x, and to use the Kalman filter to compute a more and more reliable estimation. In particular, let

$$\begin{aligned} \dot{x}(t) &= 0, \quad x(0) = \bar{x} \\ y(t) &= x(t) + v_y(t) \end{aligned}$$

Now apply the Kalman filter to this system, characterized by the matrices $A = 0$, $C = 1$, $\tilde{Q} = 0$, $\tilde{R} = r$. The Riccati equation is

$$\frac{d\tilde{P}(t)}{dt} = -\frac{\tilde{P}^2(t)}{r}$$

or

$$\frac{d\tilde{P}(t)}{\tilde{P}^2(t)} = -\frac{dt}{r}$$

Its solution is

$$\int_{\tilde{P}_0}^{\tilde{P}(t)} \frac{d\tilde{P}(t)}{\tilde{P}^2(t)} = -r^{-1} \int_0^t dt$$

or

$$-\frac{1}{\tilde{P}}\bigg|_{\tilde{P}_0}^{\tilde{P}(t)} = -r^{-1}t$$

Therefore,

$$\tilde{P}(t) = \frac{1}{r^{-1}t + \frac{1}{\tilde{P}_0}}$$

Note that the pair $\left(A, \tilde{Q}^{1/2}\right)$ is not reachable, so that one of the basic assumptions required to obtain asymptotic results is not fulfilled, and for $t \to \infty$, one has $\tilde{P}(t) \to 0$.

The gain is

$$L(t) = \tilde{P}(t)C\tilde{R}^{-1} = \frac{r^{-1}}{r^{-1}t + \frac{1}{\tilde{P}_0}}$$

and the final expression of the filter is

$$\dot{\hat{x}}(t) = \frac{r^{-1}}{r^{-1}t + \frac{1}{\tilde{P}_0}}(y(t) - \hat{x}(t))$$

Note that the gain tends to zero as $t \to \infty$, this can be explained by observing that, when new measurements of x are available, its estimate progressively improves.

Example 9.2.2. Consider the system

$$\begin{aligned} \dot{x}(t) &= Ax(t) + v_x(t) \\ y(t) &= v_y(t) \end{aligned}$$

Since $C = 0$, the gain is $L(t) = 0$ and the filter works in open loop, that is

$$\dot{\hat{x}}(t) = A\hat{x}(t)$$

In this case, the output measure y does not contain any information on the state. The only reasonable estimate can be obtained through the evolution of the system without noise starting from the initial condition x_0.

Example 9.2.3. Consider the system

$$\begin{aligned} \dot{x}(t) &= v_x(t) \\ y(t) &= Cx(t) + v_y(t) \end{aligned}$$

where the state $x \in R^1$ is a Brownian motion. Since $A = 0$, by applying the Kalman filter, the steady-state Riccati equation is

$$0 = \tilde{Q} - \frac{\bar{\tilde{P}}^2 C^2}{\tilde{R}}$$

It follows that

$$\bar{\tilde{P}} = \sqrt{\frac{\tilde{Q}\tilde{R}}{C^2}}$$

and

$$\bar{L} = \bar{\tilde{P}}C'\tilde{R}^{-1} = \sqrt{\frac{\tilde{Q}\tilde{R}}{C^2}}\frac{C}{\tilde{R}} = \sqrt{\frac{\tilde{Q}}{\tilde{R}}}$$

Hence, the Kalman filter is

$$\frac{d\hat{x}(t)}{dt} = \sqrt{\frac{\tilde{Q}}{\tilde{R}}}\,[y(t) - C\hat{x}(t)]$$

As expected, the gain depends on the ratio $\frac{\tilde{Q}}{\tilde{R}}$.

10

Linear Quadratic Gaussian control

This chapter describes the Linear Quadratic Gaussian, or *LQG*, control, obtained by combining the *LQ* solution to a deterministic problem with measurable state, and a Kalman filter used for the state estimation when stochastic noises affect the process. The relations of *LQG* with H_2 control are discussed and some hints are given on the solution to the H_∞ control problem.

10.1 *LQG* control

Consider the system

$$\begin{aligned} \dot{x}(t) &= Ax(t) + Bu(t) + v_x(t) \\ y(t) &= Cx(t) + v_y(t) \end{aligned}$$

where the noises v_x, v_y, and the initial state $x(0)$ satisfy all the assumptions introduced in Chapter 9 for the formulation and derivation of the Kalman filter. For this system, the goal is to minimize

$$J = \lim_{T\to\infty} \frac{1}{T} E\left[\int_0^T (x'(t)Qx(t) + u'(t)Ru(t))\, dt\right]$$

when the state x is nonmeasurable.

Remark. In the above formulation, since the system is affected by stochastic noises, in closed-loop x and u are stochastic processes and the expected value operator must be used.

Remark. Since v_x is a stochastic process with covariance $\tilde{Q} \geq 0$, x and u do not asymptotically vanish. Therefore, the integral term tends to infinity, irrespective of the applied control action. However, due to the term $1/T$ it is possible to obtain a finite performance index, as discussed in the following.

Remark. Under suitable assumptions, the *LQG* control law guarantees asymptotic stability of the closed-loop system, so that x and u are stationary stochastic processes. Therefore,

$$J = \lim_{T\to\infty} \frac{1}{T} E\left[\int_0^T (x'(t)Qx(t) + u'(t)Ru(t))\, dt\right]$$

can be written as

$$J = E\left[x'(t)Qx(t) + u'(t)Ru(t)\right]$$

10.1.1 Problem solution

In order to solve the *LQG* problem, first assume that all the hypotheses required for the solution to the infinite horizon *LQ* control are fulfilled, i.e. the reachability of the pair (A, B) and the observability of the pair (A, C_q), where $Q = C_q'C_q$. Assume also that the hypotheses of the stationary Kalman predictor are satisfied, i.e. the observability of the pair (A, C) and the reachability of the pair (A, B_q), where $\tilde{Q} = B_q B_q'$. It is then possible to use a "separation theorem", here not proven, which allows one to solve the *LQG* problem according to the following steps.

1. Apply the asymptotic Kalman filter to compute the observer gain $\bar{L}$ and the corresponding optimal state estimation $\hat{x}(t)$.

2. Compute the control gain $\bar{K}$ by solving the optimal *LQ* control problem for the undisturbed deterministic system with measurable state.

3. Finally, apply the control law

$$u(t) = -\bar{K}\hat{x}(t)$$

The corresponding control system is shown in Figure 10.1.

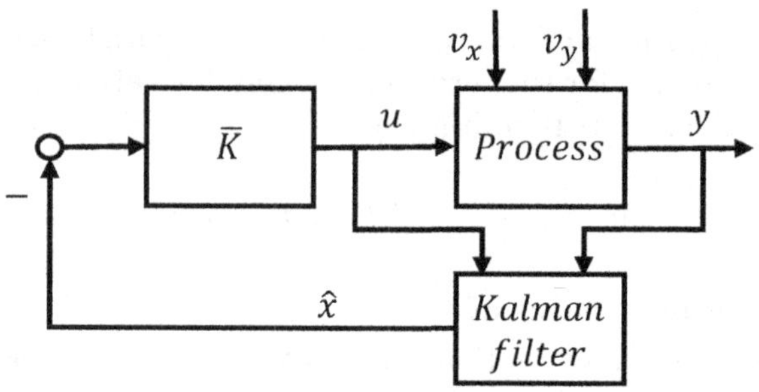

Figure 10.1: *LQG* control scheme.

The overall regulator is described by

$$\begin{aligned} \dot{\hat{x}}(t) &= A\hat{x}(t) + Bu(t) + \bar{L}\left[y(t) - C\hat{x}(t)\right] \\ u(t) &= -\bar{K}\hat{x}(t) \end{aligned}$$

or, in Laplace transforms and assuming $\hat{x}(0) = 0$,

$$s\hat{X}(s) = \left(A - \bar{L}C\right)\hat{X}(s) - B\bar{K}\hat{X}(s) + \bar{L}Y(s)$$

$$\hat{X}(s) = \left(sI - \left(A - \bar{L}C - B\bar{K}\right)\right)^{-1} \bar{L}Y(s)$$

Therefore

$$\begin{aligned} U(s) &= -\bar{K}\left(sI - \left(A - \bar{L}C - B\bar{K}\right)\right)^{-1} \bar{L}Y(s) \\ &= -R(s)Y(s) \end{aligned}$$

where

$$R(s) = \bar{K}\left(sI - \left(A - \bar{L}C - B\bar{K}\right)\right)^{-1} \bar{L}.$$

Finally, according to analogous developments reported in Chapter 6 for pole assignment control, it is easy to show that the eigenvalues of the resulting closed-loop system are those of $(A-B\bar{K})$ and of $(A-\bar{L}C)$, so that the closed-loop stability is guaranteed.

10.1.2 Robustness of *LQG* control

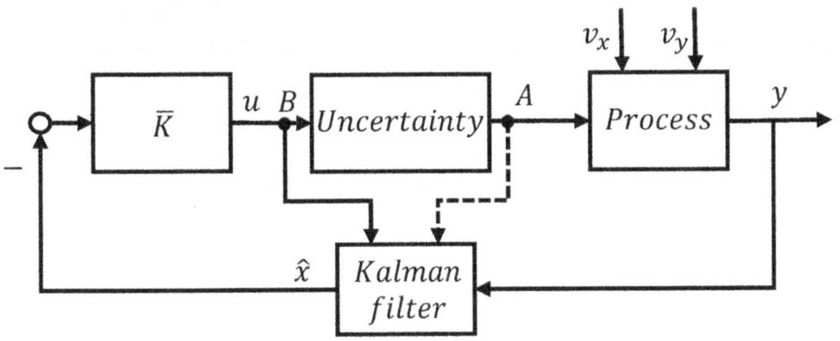

Figure 10.2: Robustness analysis in *LQG* control.

It has already been shown that the *LQ* control law guarantees robustness for phase variations of $\mp 60^0$ or gain variations $\left(\frac{1}{2}, \infty\right)$. These properties are lost with *LQG* control, as it can be qualitatively understood by noting that in the *LQ* control scheme the admissible uncertainty is at the process input. On the contrary, in *LQG* control, the scheme to be considered in shown in Figure 10.2. It is apparent that the input of the observer is the signal at point **B**, while the nominal process model is used in the design (point **A**), so that the state estimate $\hat{x}$ is in general wrong. The original *LQ* robustness could only be guaranteed by feeding the observer with the signal at point **A** of the scheme, which is unavailable. A design procedure to recover the *LQ* robustness properties is described in the following paragraph. However, note that the *LQG* regulator guarantees better attenuation properties of the high frequency measurement noise than the *LQ* controller. In fact, recall that for single input systems ($m = 1$) it has been shown that the Bode magnitude diagram of the loop transfer function with *LQ* control, i.e. $L(s) = \bar{K}(sI - A)^{-1}B$, has a shape equal to -1 at high frequency ($\omega \to \infty$). On the contrary, the *LQG* regulator is described by the strictly proper transfer function

$$R(s) = \bar{K}(sI - (A - LC - B\bar{K}))^{-1}\bar{L}$$

Since also the process transfer function

$$G(s) = C(sI - A)^{-1}B$$

is strictly proper, the corresponding loop transfer function $L(s) = G(s)R(s)$ has relative degree at least equal to 2. This means that the corresponding Bode magnitude diagram has a shape at least equal to -2 at high frequency, so guaranteeing better attenuation properties.

10.1.3 Loop Transfer Recovery (*LTR*)

In order to recover the robustness properties of the *LQ* control law also when an observer must be used, the so-called Loop Transfer Recovery (*LTR*) design procedure is adopted. This approach is now described with reference to *SISO* systems.
Consider the *SISO* system

$$\begin{aligned} \dot{x}(t) &= Ax(t) + Bu(t) \\ y(t) &= Cx(t) \end{aligned}$$

with transfer function

$$G(s) = C(sI - A)^{-1}B = \frac{B_G(s)}{A_G(s)} = \frac{b_{n-1}s^{n-1} + ... + b_0}{s^n + a_{n-1}s^{n-1} + ... + a_0}$$

First assume to be in ideal conditions, with a measurable state, and consider the state-feedback control law

$$u(t) = -Kx(t) \quad , \quad K = \left[\begin{array}{ccccc} k_0 & k_1 & \dots & k_{n-2} & k_{n-1} \end{array} \right]$$

Correspondingly, the "ideal "loop transfer function, which should be obtained also when a state observer is used, is

$$L_a^1(s) = K(sI - A)^{-1}B = \frac{\kappa(s)}{A_G(s)} \quad , \quad \kappa(s) = k_{n-1}s^{n-1} + ... + k_0$$

When the state is nonmeasurable, an observer is required and the overall control law is described by

$$\begin{aligned} \dot{\hat{x}}(t) &= A\hat{x}(t) + Bu(t) + L[y(t) - C\hat{x}(t)] \\ u(t) &= -K\hat{x}(t) \end{aligned}$$

The corresponding regulator transfer function is $R(s) = K(sI - A + BK + LC)^{-1}L$. In this case, the loop transfer function is

$$L_a^2(s) = R(s)G(s) = K(sI - A + BK + LC)^{-1}LC(sI - A)^{-1}B$$

Then, the *LTR* goal is to design the gains K and/or L such that $L_a^2(j\omega) \to L_a^1(j\omega)$ in the largest frequency band. To this end, consider the observer's gain

$$L = \rho B$$

and assume that the system is in controllable canonical form, so that

$$A = \begin{bmatrix} 0 & 1 & 0 & \dots & 0 \\ 0 & 0 & 1 & \dots & 0 \\ \dots & \dots & \dots & \dots & \dots \\ 0 & 0 & 0 & \dots & 1 \\ -a_0 & -a_1 & -a_2 & -a_0 & -a_{n-1} \end{bmatrix}, \quad B = \begin{bmatrix} 0 \\ 0 \\ \dots \\ 0 \\ 1 \end{bmatrix},$$

$$C = \begin{bmatrix} b_0 & b_1 & \dots & b_{n-2} & b_{n-1} \end{bmatrix}$$

and

$$L = \rho B = \begin{bmatrix} 0 \\ 0 \\ \dots \\ 0 \\ \rho \end{bmatrix}$$

Now consider the terms BK and ρBC appearing in the definition of the regulator $R(s)$. Both of them have the following structure

$$\begin{bmatrix} 0 & 0 & 0 & \dots & 0 \\ 0 & 0 & 0 & \dots & 0 \\ \dots & \dots & \dots & \dots & \dots \\ 0 & 0 & 0 & \dots & 0 \\ * & * & * & * & * \end{bmatrix}$$

where $*$ are generic nonzero terms. Therefore, the matrix $A - BK - \rho BC$ has the structure of a dynamic matrix in controllable canonical form, and in view of the structure of K and L, it follows that

$$R(s) = \frac{\rho\kappa(s)}{A_G(s) + \kappa(s) + \rho B_G(s)}$$

so that

$$L_a^2(s) = \frac{\rho\kappa(s)}{(A_G(s) + \kappa(s) + \rho B_G(s))} \frac{B_G(s)}{A_G(s)}$$

It then follows that, when $\rho \to \infty$

$$L_a^2(s) \to L_a^1(s) = \frac{\kappa(s)}{A_G(s)}$$

Note that in this case ($\rho \to \infty$) there is a cancellation of the open-loop zeros (polynomial B_G), which therefore must have negative real part.
In terms of the observer's eigenvalues, the requirement of stable zeros can be easily justified. In fact, it is easy to see that the observer can be written (in Laplace transforms) as

$$\hat{X}(s) = (sI - A + LC)^{-1}[BU(s) + LY(s)]$$

and

$$det(sI - A + LC) = det(sI - A + \rho BC) = A_G + \rho B_G$$

which shows that, when $\rho \to \infty$ some of the observer's eigenvalues tend to the zeros of the system (roots of B_G), while the remaining ones tend to infinity.
The previous considerations hold true also for $MIMO$ systems, and the choice $L = \rho B$ can be made to achieve the *LTR* provided that the system has the following characteristics:

1. it is square (same number of inputs and outputs),
2. its invariant zeros have negative real part.

In view of the previous considerations, the value of ρ must be increased until $L_a^2(j\omega) \simeq L_a^1(j\omega)$ in the frequency band of interest. However, note that the stability of the observer with $L = \rho B$ is not a-priori guaranteed even for systems with invariant zeros with negative real part.
An alternative approach is to obtain the loop transfer recovery with the Kalman Filter. In fact, it is possible to prove that the choice $\tilde{Q} = \alpha BB'$, $\tilde{R} = I$, with $\alpha \longrightarrow \infty$ leads to the *LTR* result. In order to give an hint, note that the associated stationary Riccati equation is

$$A\tilde{P} + \tilde{P}A' + \alpha B\tilde{R}B' - \tilde{P}C'\tilde{R}^{-1}C\tilde{P} = 0$$

or

$$A\tilde{P} + \tilde{P}A' + \alpha B\tilde{R}B' - L\tilde{R}L' = 0$$

where, for $\alpha \to \infty$, the last two terms dominate and the corresponding solution is such that $\alpha B\tilde{R}B' = L\tilde{R}L'$. This solution based on *KF*, which also requires that the invariant zeros of the system have negative real part, is better than to directly set $L = \rho B$ since the stability properties of the Kalman filter guarantee that $A - LC$ is stable for any value of α.

10.2 $LQG - H_2$ control

In Chapter 3 it has been shown how to formulate an H_2 control problem which consists of the selection of a regulator $R(s)$ minimizing the H_2 norm of the transfer function between the input w (exogenous signal) and the output z (performance variables) in the control scheme reported in Figure 10.3, where P is an enlarged plant and v are the measured variables.

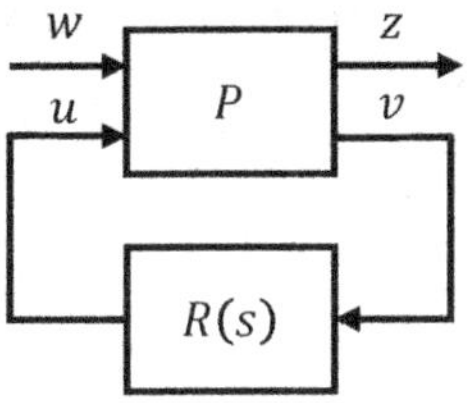

Figure 10.3: General scheme for H_2, H_∞ control.

The enlarged plant P can be given the very general state-space representation

$$\begin{aligned}\dot{x}(t) &= Ax(t) + B_1 w(t) + B_2 u(t)\\ z(t) &= C_1 x(t) + D_{11} w(t) + D_{12} u(t)\\ v(t) &= C_2 x(t) + D_{21} w(t) + D_{22} u(t)\end{aligned}$$

or, analogously,

$$P = \begin{bmatrix} A & B_1 & B_2 \\ C_1 & D_{11} & D_{12} \\ C_2 & D_{21} & D_{22} \end{bmatrix}$$

where B_1, B_2, C_1, C_2, D_{11}, D_{12}, D_{21}, D_{22} are matrices to be suitably selected to meet specific design requirements.

10.2.1 H_2 formulation of *LQG* control

In view of the previous considerations, given the system under control described by

$$\begin{aligned}\dot{x}(t) &= Ax(t) + Bu(t) + v_x(t)\\ y(t) &= Cx(t) + v_y(t)\end{aligned}$$

where v_x and v_y are zero-mean white noises with covariance $\tilde{Q}$ and $\tilde{R}$, the minimization of the *LQG* cost function

$$J = E\left[\lim_{T\to\infty}\frac{1}{T}\int_0^\infty [x'Qx + u'Ru]\,dt\right]$$

corresponds to minimize $\|G_{zw}\|_2^2$ where

$$z = \begin{bmatrix} Q^{1/2} & 0 \\ 0 & R^{1/2} \end{bmatrix}\begin{bmatrix} x \\ u \end{bmatrix}, \quad \begin{bmatrix} v_x \\ v_y \end{bmatrix} = \begin{bmatrix} \tilde{Q}^{1/2} & 0 \\ 0 & \tilde{R}^{1/2} \end{bmatrix} w$$

and w is a white noise with zero mean and unit variance. The plant model to be considered is then described by

$$P = \begin{bmatrix} A & \tilde{Q}^{1/2} & 0 & B \\ Q^{1/2} & 0 & 0 & 0 \\ 0 & 0 & 0 & R^{1/2} \\ C & 0 & \tilde{R}^{1/2} & 0 \end{bmatrix}$$

Equivalently,

$$B_1 = \begin{bmatrix} \tilde{Q}^{1/2} & 0 \end{bmatrix} \quad , \quad B_2 = B \quad , \quad C_1 = \begin{bmatrix} Q^{1/2} \\ 0 \end{bmatrix}$$

$$D_{11} = \begin{bmatrix} 0 & 0 \\ 0 & 0 \end{bmatrix} \quad , \quad D_{12} = \begin{bmatrix} 0 \\ R^{1/2} \end{bmatrix} \quad , \quad C_2 = C$$

$$D_{21} = \begin{bmatrix} 0 & \tilde{R}^{1/2} \end{bmatrix} \quad , \quad D_{22} = 0$$

This result proves that the ***LQG*** control problem is a particular case of the more general H_2 problem, whose solution is given by a static feedback applied to an estimated state computed with a Kalman filter.

10.2.2 Shaping the sensitivity functions with *LQG* - H_2 control

The main difficulties related to the use of ***LQG*** control are due to the proper selection of the state and control weighting matrices Q and R, as well as of the covariance functions $\tilde{Q}$, $\tilde{R}$ so as to guarantee that the sensitivity functions are properly weighted. To this regard, some synthesis techniques based on a reformulation of the ***LQG*** - H_2 control problem are now presented.

Shaping functions at the process inputs

Consider the control scheme reported in Figure 10.4, where v_d, v_{n1}, and v_{n2} are uncorrelated white noises and S_d, S_n are asymptotically stable dynamic systems producing the stationary noises d and n.

Assume as usual that the plant P is described by

$$\begin{aligned} \dot{x}(t) &= Ax(t) + Bu(t) \\ \tilde{z}(t) &= Cx(t) + d(t) \\ y(t) &= \tilde{z}(t) + n(t) \end{aligned}$$

while S_d and S_n are described by

$$\begin{aligned} \dot{x}_d(t) &= A_d x_d(t) + B_d v_d(t) \\ d(t) &= C_d x_d(t) \end{aligned}$$

and

$$\begin{aligned} \dot{x}_n(t) &= A_n x_n(t) + B_n v_{n1}(t) \\ n(t) &= C_n x_n(t) + v_{n2}(t) \end{aligned}$$

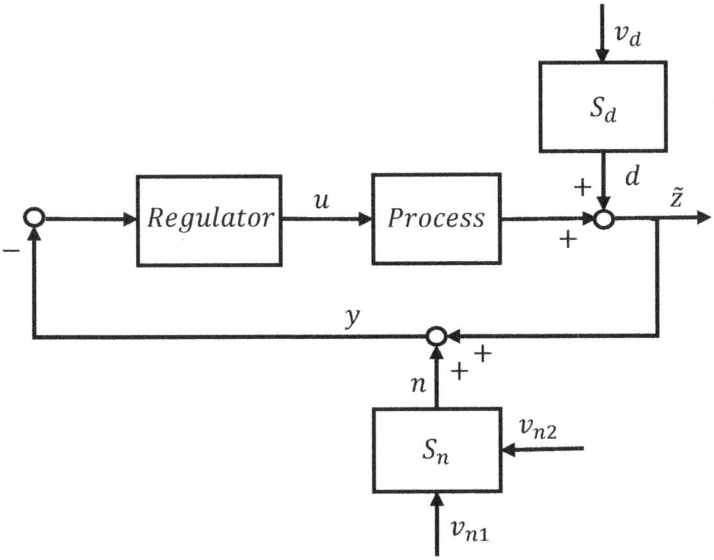

Figure 10.4: Scheme with disturbances prefiltering.

By combining the three systems, one obtains

$$
\begin{aligned}
\begin{bmatrix} \dot{x}(t) \\ \dot{x}_d(t) \\ \dot{x}_n(t) \end{bmatrix} &= \begin{bmatrix} A & 0 & 0 \\ 0 & A_d & 0 \\ 0 & 0 & A_n \end{bmatrix} \begin{bmatrix} x(t) \\ x_d(t) \\ x_n(t) \end{bmatrix} + \begin{bmatrix} B \\ 0 \\ 0 \end{bmatrix} u(t) \\
&+ \begin{bmatrix} 0 & 0 \\ B_d & 0 \\ 0 & B_n \end{bmatrix} \begin{bmatrix} v_d(t) \\ v_{n1}(t) \end{bmatrix} \\
y(t) &= \begin{bmatrix} C & C_d & C_n \end{bmatrix} \begin{bmatrix} x(t) \\ x_d(t) \\ x_n(t) \end{bmatrix} + v_{n2}(t) \\
\tilde{z}(t) &= \begin{bmatrix} C & C_d & 0 \end{bmatrix} \begin{bmatrix} x(t) \\ x_d(t) \\ x_n(t) \end{bmatrix}
\end{aligned}
$$

Now define

$$
\begin{aligned}
\bar{x}(t) &= \begin{bmatrix} x(t) \\ x_d(t) \\ x_n(t) \end{bmatrix}, \quad \bar{A} = \begin{bmatrix} A & 0 & 0 \\ 0 & A_d & 0 \\ 0 & 0 & A_n \end{bmatrix}, \quad \bar{B} = \begin{bmatrix} B \\ 0 \\ 0 \end{bmatrix} \\
\bar{v}(t) &= \begin{bmatrix} 0 & 0 \\ B_d & 0 \\ 0 & B_n \end{bmatrix} \begin{bmatrix} v_d(t) \\ v_{n1}(t) \end{bmatrix}, \quad \bar{C} = \begin{bmatrix} C & C_d & C_n \end{bmatrix}
\end{aligned}
$$

Assuming that the covariance $\tilde{Q}_{n2}$ of v_{n2} is positive definite, the enlarged system described by

$$\begin{aligned} \dot{\bar{x}}(t) &= \bar{A}\bar{x}(t) + \bar{B}u(t) + \bar{v}(t) \\ y(t) &= \bar{C}\bar{x}(t) + v_{n2}(t) \end{aligned}$$

fulfills the conditions required for the design of a stabilizing *LQG* regulator. Note that, as apparent from Figure 10.4, the systems S_d e S_n are non reachable from the input u and must be asymptotically stable.
In view of the spectral factorization theorem, d and n are stationary stochastic processes; in turn this implies that, if the feedback system is asymptotically stable, also x, u, y, and $\tilde{z}$ are stationary stochastic processes and it is possible to consider the following cost function

$$J = E\left[\tilde{z}'(t)\tilde{z}(t) + u'(t)u(t)\right]$$

Now note that, denoting by $R(s)$, $S(s)$, and $T(s)$ the regulator, the sensitivity, and the complementary sensitivity functions, respectively, one has

$$\begin{aligned} \tilde{Z}(s) &= -T(s)N(s) + S(s)D(s) \\ U(s) &= -R(s)S(s)\left[N(s) + D(s)\right] \end{aligned}$$

where $N(s)$ and $D(s)$ are the Laplace transforms of $n(t)$ and $d(t)$.
In view of the stationarity of $\tilde{z}$ and u, the following relations hold

$$E\left[\tilde{z}'(t)\tilde{z}(t)\right] = \frac{1}{2\pi}\int_{-\infty}^{\infty} tr\left\{\Phi_{\tilde{z}\tilde{z}}\left(\omega\right)\right\} d\omega$$

and

$$E\left[u'(t)u(t)\right] = \frac{1}{2\pi}\int_{-\infty}^{\infty} tr\left\{\Phi_{uu}\left(\omega\right)\right\} d\omega$$

where $\Phi_{\tilde{z}\tilde{z}}$ and Φ_{uu} are the power spectral densities of z and u.
In view of the previous results, one has

$$\begin{aligned} tr\left\{\Phi_{\tilde{z}\tilde{z}}\left(\omega\right)\right\} &= tr\left\{T(j\omega)\Phi_{nn}\left(\omega\right) T'(-j\omega)\right\} + \\ &\quad + tr\left\{S(j\omega)\Phi_{dd}\left(\omega\right) S'(-j\omega)\right\} \end{aligned}$$

and

$$tr\left\{\Phi_{uu}\left(\omega\right)\right\} = tr\left\{R(j\omega)S(j\omega)\left(\Phi_{nn}\left(\omega\right) + \Phi_{dd}\left(\omega\right)\right) S'(-j\omega)R'(-j\omega)\right\}$$

The sum of these two elements corresponds to

$$tr\left\{\Gamma(j\omega)\Gamma'(-j\omega)\right\}$$

where

$$\Gamma(j\omega) = \begin{bmatrix} T(j\omega)\Phi_{nn}^{1/2}\left(\omega\right) \\ S(j\omega)\Phi_{dd}^{1/2}\left(\omega\right) \\ R(j\omega)S(j\omega)\left(\Phi_{nn}\left(\omega\right) + \Phi_{dd}\left(\omega\right)\right)^{1/2} \end{bmatrix}$$

so that, in conclusion,

$$J = \frac{1}{2\pi}\int_{-\infty}^{\infty} tr\left\{\Gamma(j\omega)\Gamma'(-j\omega)\right\} d\omega = \|\Gamma(j\omega)\|_2^2$$

This expression, shows that a proper selection of the power spectral densities $\Phi_{nn}(\omega)$ and $\Phi_{dd}(\omega)$, i.e. of the systems S_n and S_d, allows one to weight, in the ***LQG*** problem formulation, the sensitivity functions $S(s)$, $T(s)$, and $R(s)S(s)$. This approach can be seen as the solution to a standard H_2 control problem with input shaping functions defined by the systems S_n and S_d, as shown in Figure 10.5 and the inputs, outputs, and matrices of the general H_2 formulation for system

$$P = \begin{bmatrix} A & B_1 & B_2 \\ C_1 & D_{11} & D_{12} \\ C_2 & D_{21} & D_{22} \end{bmatrix}$$

are $w = \begin{bmatrix} v_d \\ v_{n1} \\ v_{n2} \end{bmatrix}$, $z = \begin{bmatrix} \tilde{z} \\ u \end{bmatrix}$, $A = \bar{A}$, $B_2 = \bar{B}$, $B_1 = \begin{bmatrix} 0 & 0 & 0 \\ B_d & 0 & 0 \\ 0 & B_n & 0 \end{bmatrix}$, $C_1 = \begin{bmatrix} C & C_d & 0 \\ 0 & 0 & 0 \end{bmatrix}$, $C_2 = \bar{C}$, $D_{21} = \begin{bmatrix} 0 & 0 & I \end{bmatrix}$, $D_{12} = \begin{bmatrix} 0 \\ I \end{bmatrix}$, while D_{11} and D_{22} are null matrices of proper dimensions.

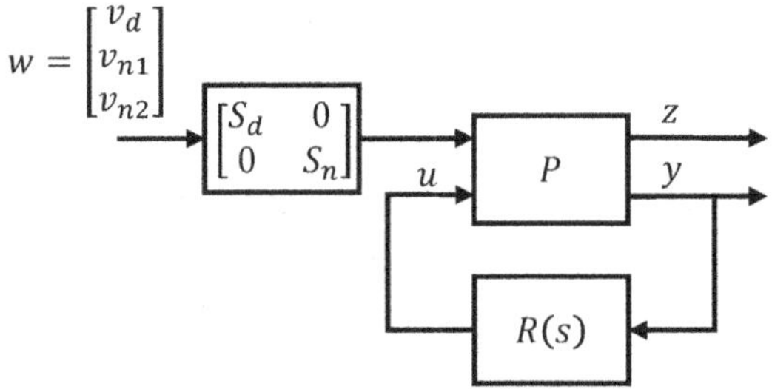

Figure 10.5: Scheme input shaping functions.

Shaping functions at the process outputs

It is more frequent to consider suitable shaping function at the output of the system, according to the scheme reported in Figure 10.6, where $W_S(s)$, $W_K(s)$, and $W_T(s)$ are transfer functions of asymptotically stable systems, which can be viewed as the weighting functions of the system's sensitivities. An equivalent block diagram representation is given in Figure 10.7. In fact, note that

$$\begin{aligned} Z_S(s) &= W_S(s)S(s)W(s), \quad Z_K(s) = W_K(s)R(s)S(s)W(s), \\ Z_T(s) &= W_T(s)T(s)W(s) \end{aligned}$$

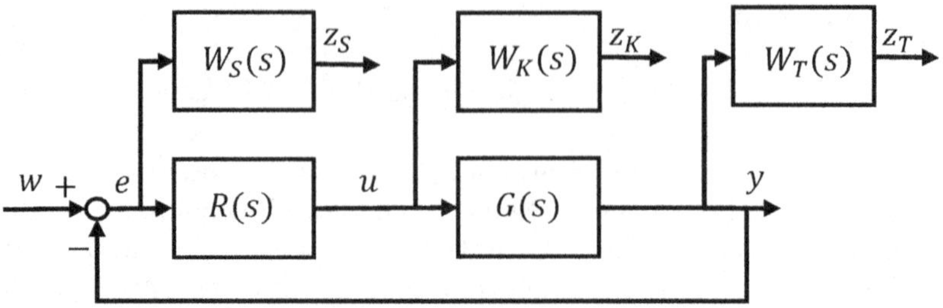

Figure 10.6: Scheme with output shaping functions.

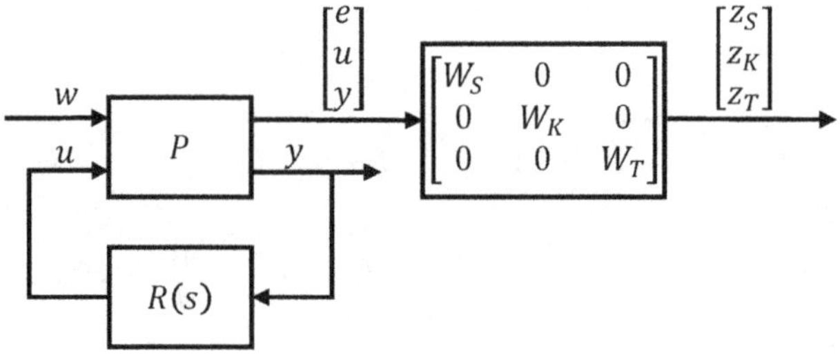

Figure 10.7: Alternative scheme with output shaping functions.

Let the systems $W_S(s)$, $W_K(s)$, $W_T(s)$ be described in state space form by the following expressions

$W_S(s)$

$$\begin{aligned} \dot{x}_S(t) &= A_S x_S(t) + B_S(w(t) - y(t)) \\ z_S(t) &= C_S x_S(t) \end{aligned}$$

$W_K(s)$

$$\begin{aligned} \dot{x}_K(t) &= A_K x_K(t) + B_K u(t) \\ z_K(t) &= C_K x_K(t) + D_K u(t) \end{aligned}$$

$W_T(s)$

$$\begin{aligned} \dot{x}_T(t) &= A_T x_T(t) + B_T y(t) \\ z_T(t) &= C_T x_T(t) + D_T y(t) \end{aligned}$$

and, in compact form suitable for the *LQG* or H_2 formulations,

$$\begin{bmatrix} \dot{x}(t) \\ \dot{x}_S(t) \\ \dot{x}_K(t) \\ \dot{x}_T(t) \end{bmatrix} = \begin{bmatrix} A & 0 & 0 & 0 \\ -B_S C & A_S & 0 & 0 \\ 0 & 0 & A_K & 0 \\ B_T C & 0 & 0 & A_T \end{bmatrix} \begin{bmatrix} x(t) \\ x_S(t) \\ x_K(t) \\ x_T(t) \end{bmatrix} + \begin{bmatrix} B \\ 0 \\ B_K \\ 0 \end{bmatrix} u(t) + \begin{bmatrix} 0 \\ B_S \\ 0 \\ 0 \end{bmatrix} w(t)$$

$$\begin{bmatrix} z_S(t) \\ z_K(t) \\ z_T(t) \end{bmatrix} = \begin{bmatrix} 0 & C_S & 0 & 0 \\ 0 & 0 & C_K & 0 \\ D_T C & 0 & 0 & C_T \end{bmatrix} \begin{bmatrix} x(t) \\ x_S(t) \\ x_T(t) \\ x_K(t) \end{bmatrix} + \begin{bmatrix} 0 \\ D_K \\ 0 \end{bmatrix} u(t) + \begin{bmatrix} 0 \\ 0 \\ 0 \end{bmatrix} w(t)$$

$$v(t) = \begin{bmatrix} -C & 0 & 0 & 0 \end{bmatrix} \begin{bmatrix} x(t) \\ x_S(t) \\ x_K(t) \\ x_T(t) \end{bmatrix} + w(t)$$

Note in particular that $D_{11} = 0$ and $D_{22} = 0$.

10.2.3 Solution to the H_2 and H_∞ control problems

The general solutions of the H_2 and H_∞ control problems are now given under the following assumptions

1. The triple (A, B_2, C_2) is stabilizable and detecable;
2. The triple (A, B_1, C_1) is stabilizable and detectable;
3. $D_{11} = 0$ and $D_{22} = 0$;
4. $D_{12}^{'} C_1 = 0$;
5. $B_1 D_{21}^{'} = 0$;
6. $D_{12}' D_{12} = I$;
7. $D_{21}' D_{21} = I$;
8. $\begin{bmatrix} A - j\omega I & B_2 \\ C_1 & D_{12} \end{bmatrix}$ has full column rank $\forall \omega$;
9. $\begin{bmatrix} A - j\omega I & B_1 \\ C_2 & D_{21} \end{bmatrix}$ has full row rank $\forall \omega$.

Some of these assumptions are mandatory to guarantee the problem solution, while some others are introduced just for simplicity (depending also on the specific problem considered, i.e. H_2 or H_∞). Note also that, with reference to the standard *LQG* control problem, the assumptions $D_{11} = 0$, $D_{22} = 0$, $D_{12}'C_1 = 0$, $B_1 D_{21}' = 0$ are always verified, while $D_{12}'D_{12} = I$ and $D_{21}'D_{21} = I$ can be guaranteed provided that the control weighting matrix is chosen as $R = I$ and the covariance of the output noise is $\tilde{R} = I$. This is not limiting indeed, since it is always possible to recover this case by a proper scaling of the original parameters.

Solution to the H_2 control problem

The H_2 controller is given by the standard state feedback control law

$$u(t) = -K\hat{x}(t) \quad , \quad K = B_2'P$$

where P is the unique positive definite solution to the Riccati equation

$$A'P + PA - PB_2B_2'P + C_1'C_1 = 0$$

and by the observer's equation

$$\dot{\hat{x}}(t) = A\hat{x}(t) + B_2u(t) + L(y(t) - C_2\hat{x}(t))$$

where the gain L is

$$L = \tilde{P}C_2'$$

and $\tilde{P}$ is the unique positive definite solution to the Riccati equation

$$A\tilde{P} + \tilde{P}A - \tilde{P}C_2C_2\tilde{P} + B_1B_1' = 0$$

Solution to the H_∞ control problem

The H_∞, control problem requires to design a regulator guaranteeing a given attenuation level γ, i.e. such that

$$\|G_{zw}\|_\infty < \gamma$$

Under the above assumptions, a regulator guaranteeing the attenuation level γ exists if and only if

1. $P \geq 0$ is a solution to the algebraic Riccati equation

$$A'P + PA + C_1'C_1 + P\left(\gamma^{-2}B_1B_1' - B_2B_2'\right)P = 0$$

   such that

   $Re(\lambda_i[A + (\gamma^{-2}B_1B_1' - B_2B_2')P]) < 0$ for any i

2. $S \geq 0$ is a solution to the algebraic Riccati equation

$$AS + SA' + B_1B_1' + S\left(\gamma^{-2}C_1'C_1 - C_2'C_2\right)S = 0$$

such that

$Re(\lambda_i[A + S(\gamma^{-2}C_1'C_1 - C_2'C_2)]) < 0$ for any i

and

3.

$$\max_i |\lambda_i(PS)| < \gamma^2$$

Also in this case, the regulator is made by a state observer and an algebraic control law

$$\begin{aligned} \dot{\hat{x}}(t) &= A\hat{x}(t) + \gamma^{-2}B_1B_1'P\hat{x}(t) + B_2u(t) + ZL\left(y(t) - C_2\hat{x}(t)\right) \\ u(t) &= -K\hat{x}(t) \end{aligned}$$

where

$$Z = \left(I - \gamma^{-2}SP\right)^{-1}, \ L = SC_2', \ K = B_2'P$$

Note that the minimum value of γ is not a priori known, and an iterative procedure must be used. In practice, the value of γ is progressively reduced until the minimum value satisfying the previous relations is met.

In order to verify the result above, at least in a simple case, assume that the state is known and the system is described by

$$\begin{aligned} \dot{x}(t) &= Ax(t) + B_1w(t) + B_2u(t) \\ z(t) &= C_1x(t) + D_{12}u(t) \end{aligned}$$

Letting $x(0) = 0$, we want to verify that the control law

$$u(t) = -B_2'Px$$

guarantees the attenuation level γ from w to z, that is

$$\|z\|_2 \leq \gamma \|w\|_2$$

Consider the function

$$V(t) = x'(t)Px(t) + \int_0^t (z'(\tau)z(\tau) - \gamma^2 w'(\tau)w(\tau))d\tau$$

It is apparent that, if $V(t) \leq 0$ for any t, the integrand is negative and the prescribed attenuation level is attained.

Note preliminarily that $V(0) = 0$, then, if $\dot{V}(t) \leq 0$ for any t, V is negative or null at any time. But, removing the dependence on time for readability,

$$\begin{aligned} \dot{V} &= \dot{x}'Px + x'P\dot{x} + z'z - \gamma^2 w'w \\ &= (Ax + B_1w + B_2u)'Px + x'P(Ax + B_1w + B_2u) \\ &\quad +(C_1x + D_{12}u)'(C_1x + D_{12}u) - \gamma^2 w'w \\ &= x'(A'P + PA + C_1'C_1)x + (u' + x'PB_2)(u + B_2'Px) \\ &\quad -x'PB_2B_2'Px + w'B_1'Px + x'PB_1w - \gamma^2 w'w \\ &= (a) \end{aligned}$$

where it has been summed and subtracted $x'PB_2B_2'Px$. Again, by summing and subtracting $\frac{1}{\gamma^2}PB_1B_1'P$ it is possible to write

$$
\begin{aligned}
(a) &= x'(A'P + PA + C_1'C_1 - PB_2B_2'P + \frac{1}{\gamma^2}PB_1B_1'P)x \\
&\quad +(u' + x'PB_2)(u + B_2'Px) - \gamma^2(w' - \frac{1}{\gamma^2}x'PB_1)(w - \frac{1}{\gamma^2}B_1'Px)
\end{aligned}
$$

Therefore, if P is computed as the solution to the previous Riccati equation and the control law $u(t) = -B_2'Px$ is used, one obtains

$$\dot{V} = -\gamma^2(w' - \frac{1}{\gamma^2}x'PB_1)(w - \frac{1}{\gamma^2}B_1'Px) \leq 0$$

and the result is proven. Note that the worst disturbance is

$$w = \frac{1}{\gamma^2}B_1'Px$$

10.3 Model reduction techniques

The H_2 and H_∞ approaches lead to the design of a regulator, made by a static feedback control law and by an observer, with order equal to the order of the plant, plus the order of the adopted shaping functions W_S, W_T, and W_K. This order can be very high, and it is advisable to use model order reduction techniques to compute a simpler approximate regulator.

To this end, consider the system

$$
\begin{aligned}
\dot{x}(t) &= Ax(t) + Bu(t) \\
y(t) &= Cx(t) + Du(t)
\end{aligned}
$$

Its *controllability gramian* P and *observability gramian* Q are defined as

$$
\begin{aligned}
P &= \int_0^\infty e^{At}BB'^{A't}dt \\
Q &= \int_0^\infty e^{A't}C'Ce^{At}dt
\end{aligned}
$$

Alternatively, for asymptotically stable systems, P and Q are the positive definite solutions of

$$
\begin{aligned}
AP + PA' + BB' &= 0 \\
A'Q + QA + C'C &= 0
\end{aligned}
$$

In general, $P \neq Q$, however, for reachable and observable systems it is possible to find a state transformation such that $P = Q$, in this case the system is in the so-called *balanced realization* and

$$P = Q = \Sigma = diag(\sigma_1, \sigma_2, \ldots, \sigma_n), \; \sigma_1 \geq \sigma_2 \geq \ldots \geq \sigma_n > 0 \qquad (10.1)$$

Given the system's balanced realization, it is possible to reduce the order of the system by removing the states corresponding to the smaller $\sigma_i' s$. Assuming to consider only the first k values $\sigma_1, \sigma_2, \ldots, \sigma_k$, the (balanced) system can be partitioned as

$$A = \begin{bmatrix} A_{11} & A_{12} \\ A_{21} & A_{22} \end{bmatrix}, B = \begin{bmatrix} B_1 \\ B_2 \end{bmatrix}, C = \begin{bmatrix} C_1 & C_2 \end{bmatrix}$$

where $A_{11} \in R^{k,k}$, $B_1 \in R^{k,m}$, $C_1 \in R^{p,k}$, and Σ is given by

$$\Sigma = \begin{bmatrix} \Sigma_1 & 0 \\ 0 & \Sigma_2 \end{bmatrix}$$

with $\Sigma_1 \in R^{k,k}$. Different model reduction techniques can now be used.

Balanced Truncation

The reduced model is described by (A_{11}, B_1, C_1, D). Denoting by $G_a^k(s)$ the transfer function of this reduced model and by $G(s)$ the one of the original system, it is possible to show that

$$\|G(s) - G_a^k(s)\|_\infty \leq 2(\sigma_{k+1} + \sigma_{k+2} + \ldots + \sigma_{k+n})$$

Balanced Residualization

In this case it is assumed that $\dot{x}_2 = 0$ (i.e. the dynamics of the states to be neglected is ignored). Moreover, assuming that A_{22} is nonsingular, the reduced order model is

$$\Sigma : \begin{cases} \dot{x}_1(t) = A_r x_1(t) + B_r u(t) \\ y(t) = C_r x_1(t) + D_r u(t) \end{cases}$$

where

$$\begin{aligned} A_r = A_{11} - A_{12}A_{22}^{-1}A_{21},\ B_r = B_1 - A_{12}A_{22}^{-1}B_2 \\ C_r = C_1 - C_2A_{22}^{-1}A_{21},\ D_r = D - C_2A_{22}^{-1}B_2 \end{aligned}$$

Also in this case it can be shown that

$$\|G(s) - G_a^k(s)\|_\infty \leq 2(\sigma_{k+1} + \sigma_{k+2} + \ldots + \sigma_{k+n})$$

Moreover, the balanced residualization guarantees that $G_a^k(0) = G(0)$, so that the original static gain is maintained.

11

Optimal control of discrete time systems

In this chapter the results previously derived for continuous time systems on optimal control, Linear Quadratic (*LQ*) control, Kalman filtering (*KF*) and Linear Quadratic Gaussian control (*LQG*) are extended to discrete time systems. Since there are no significant differences with respect to the continuous time case, the presentation will be very synthetic, focusing only on some peculiarities of the considered case.

11.1 Optimal control

Consider the nonlinear, discrete-time, time-invariant system described by

$$x(k+1) = f(x(k), u(k)), \quad x(k_0) = x_0$$

with $x \in R^n$, $u \in R^m$. Assuming that the state is measurable, the optimal control problem consists of minimizing, at the initial time k_0, and with respect to the future control sequence $u(k_0)$, $u(k_0+1)$, ..., $u(\bar{k}-1)$, the cost function

$$J(x(k_0), u(\cdot), k_0) = \sum_{i=k_0}^{\bar{k}-1} l(x(i), u(i)) + m(x(\bar{k})) \quad , \quad \bar{k} > k_0$$

subject to the constraints due to the system's dynamics.

Denote by $u_{[a,b]}$ the control sequence $u(k)$, $k \in [a, b]$, and define

$$J^0(x(k), k) = \min_{u[k,\bar{k}-1]} J(x(k), u(\cdot), k)$$

then

$$J^0(x(\bar{k}), \bar{k}) = \min_{u(\bar{k})} m(x(\bar{k})) = m(x(\bar{k}))$$

and

$$J^0(x(\bar{k}-1), \bar{k}-1) = \min_{u(\bar{k}-1)} \left\{ l(x(\bar{k}-1), u(\bar{k}-1)) + J^0(x(\bar{k}), \bar{k}) \right\}$$

In general

$$J^0(x(k), k) = \min_{u(k)} \left\{ l(x(k), u(k)) + J^0(f(x(k), u(k)), k+1) \right\}$$

Eliminating the time dependence if not required, for the same reasons discussed for continuous time systems, the Hamilton- Jacobi - Bellman equation becomes

$$J^0(x,k) = \min_u \left\{ l(x,u) + J^0(f(x,u), k+1) \right\}$$

with boundary condition

$$J^0(x,\bar{k}) = m(x)$$

Also in this case, in order to find the solution to the optimization problem it is necessary, at least conceptually, to proceed according to the following steps

1. compute the value u^o minimizing

   $$\left\{ l(x,u) + J^0(f(x,u), k+1) \right\}$$

   Assuming that there exists an unique minimum, this corresponds to compute a function with arguments x and $J^0(f(x,u), k+1)$, that is

   $$u^o = \kappa\left(x, J^0\right)$$

2. Compute the function $J^o(x,k)$ satisfying the HJB equation

   $$J^o(x,k) = l\left(x, \kappa\left(x, J^o(x,k)\right)\right) + J^0\left(f(x, \kappa\left(x, J^o(x,k)\right)), k+1\right)$$

   with boundary condition

   $$J^0(x,\bar{k}) = m(x)$$

11.2 Linear Quadratic - *LQ* control

Assume that the system under control is linear

$$x(k+1) = Ax(k) + Bu(k)$$

the cost function to be minimized is quadratic

$$l(x,u) = x'Qx + u'Ru, \quad Q \geq 0, \quad R > 0$$

as well as the terminal cost

$$m(x) = x'Sx, \quad S \geq 0$$

A tentative expression for the optimal cost can be

$$J^0(x,k) = x'P(k)x$$

with $P(\bar{k}) = S$. Following the steps considered above, the optimal control law can be obtained by minimizing

$$\begin{aligned}
& l(x,u) + J^0(f(x,u), k+1) \\
= \;& x'Qx + u'Ru + (Ax+Bu)'P(k+1)(Ax+Bu) \\
= \;& x'Qx + u'Ru + x'A'P(k+1)Ax + x'A'P(k+1)Bu \\
& +u'B'P(k+1)Ax + u'B'P(k+1)Bu \\
= \;& x'(Q + A'P(k+1)A)x + u'(R + B'P(k+1)B)u \\
& +2x'A'P(k+1)Bu
\end{aligned}$$

Differentiating with respect to u and equating to zero, it follows

$$2u'(R + B'P(k+1)B) = -2x'A'P(k+1)B$$

then

$$\begin{aligned}
u(k) &= -\left(R + (B'P'(k+1)B)'\right)^{-1} B'P'(k+1)Ax(k) \\
&= -K(k)x(k)
\end{aligned}$$

with

$$K(k) = (R + B'P'(k+1)B)^{-1} B'P'(k+1)A$$

Substituting the expression of $u(k)$ in the HJB equation and letting $P = P'$, it follows

$$\begin{aligned}
& x'P(k)x \\
= \;& x'(Q + A'P(k+1)A)x \\
& +x'A'P(k+1)B\left(R + B'P(k+1)B\right)^{-1} B'P(k+1)Ax \\
& -2x'A'P(k+1)B\left(R + B'P(k+1)B\right)^{-1} B'P(k+1)Ax \\
= \;& x'\left[Q + A'P(k+1)A - A'P(k+1)B\left(R + B'P(k+1)B\right)^{-1} BP(k+1)A\right] x
\end{aligned}$$

In conclusion, the Riccati equation for the updating of $P(k)$ is

$$P(k) = Q + A'P(k+1)A - A'P(k+1)B\left(R + B'P(k+1)B\right)^{-1} B'P(k+1)A$$

with initial condition

$$P(\bar{k}) = S$$

and the control law is

$$u(k) = -K(k)x(k)$$

Note that also in this case the control law is a function of the current state, that is a feedback control law.

11.2.1 Infinite horizon LQ control

If an infinite horizon cost function is considered

$$J = \sum_{k=0}^{\infty} x'(k)Qx(k) + u'(k)Ru(k), \quad Q \geq 0, \quad R > 0$$

the following theorem can be proven.

Theorem 11.2.1. *If*

1. *the pair (A, B) is reachable;*
2. *the pair (A, C_q) is observable, with C_q such that $Q = C_q' C_q$*

then

A) *the optimal control law is given by*

$$u(k) = -\bar{K}x(k)$$

with

$$\bar{K} = \left(R + B'\bar{P}B\right)^{-1} B'\bar{P}A$$

where $\bar{P}$ is the unique positive definite solution to the stationary Riccati equation

$$\bar{P} = A'\bar{P}A + Q - A'\bar{P}B\left(R + B'\bar{P}B\right)^{-1} B'\bar{P}A$$

B) *the closed-loop system*

$$x(k+1) = \left(A - B\bar{K}\right)x(k)$$

is asymptotically stable.

Infinite horizon *LQ* robustness

For simplicity of notation, denote by P the solution to the algebraic Riccati equation associated with the infinite horizon LQ problem

$$P = A'PA + Q - A'PB\left(R + B'PB\right)^{-1} B'PA$$

and by K the associated gain

$$K = \left(R + B'PB\right)^{-1} B'PA$$

By means of simple computations, the Riccati equation can be rewritten in the following way

$$P = A'PA + Q - K'\left(R + B'PB\right)K$$

This equation is equivalent to

$$\begin{aligned}&\left(z^{-1}I - A'\right) P\left(zI - A\right)\\ = &-\left(z^{-1}I - A'\right) PA - A'P\left(zI - A\right) + Q - K'\left(R + B'PB\right)K\end{aligned}$$

By multiplying from the left by $B'\left(z^{-1}I - A'\right)^{-1}$ and from the right by $\left(zI - A\right)^{-1}B$ the previous expression becomes

$$\begin{aligned}B'PB &= -B'PA(zI - A)^{-1}B - B'(z^{-1}I - A')^{-1}A'PB\\ &\quad +B'(z^{-1}I - A')^{-1}Q(zI - A)^{-1}B\\ &\quad -B'(z^{-1}I - A')^{-1}K'\left(R + B'PB\right)K(zI - A)^{-1}B\end{aligned}$$

But

$$\begin{aligned}B'PA &= \left(R + B'PB\right)K\\ A'PB &= K'\left(R + B'PB\right)\end{aligned}$$

hence

$$\begin{aligned}B'PB &= -\left(R + B'PB\right)K\left(zI - A\right)^{-1}B\\ &\quad -B'\left(z^{-1}I - A'\right)^{-1}K'\left(R + B'PB\right)\\ &\quad +B'\left(z^{-1}I - A'\right)^{-1}Q\left(zI - A\right)^{-1}B\\ &\quad -B'\left(z^{-1}I - A'\right)^{-1}K'\left(R + B'PB\right)K\left(zI - A\right)^{-1}B\end{aligned}$$

By adding to both terms R and rearranging one obtains

$$\begin{aligned}&\left(I + B'\left(z^{-1}I - A'\right)^{-1}K'\right)\left(R + B'PB\right)\left(I + K\left(zI - A\right)^{-1}B\right)\\ = \;& R + B'\left(z^{-1}I - A'\right)^{-1}Q\left(zI - A\right)^{-1}B\end{aligned}$$

In the case of systems with a single input ($m = 1$) from the previous expression it follows

$$\left|1 + K\left(zI - A\right)^{-1}B\right|^2 \geq \frac{R}{R + B'PB} = \bar{\gamma}$$

with $\bar{\gamma} < 1$. Note now that $L(z) = K\left(zI - A\right)^{-1}B$ is the loop transfer function of the system reported in Figure 11.1.

This result shows that the polar diagram of $L\left(e^{j\theta}\right)$, $\theta \in [0, \pi]$, does not intersect the circle with center in -1 and radius $\gamma = \bar{\gamma}^{1/2}$, as shown in Figure 11.2. Then, it is possible to conclude that the gain margin is between $\frac{1}{1+\gamma}$ and $\frac{1}{1-\gamma}$ and it is less than the one guaranteed by the *LQ* control of continuous time systems.

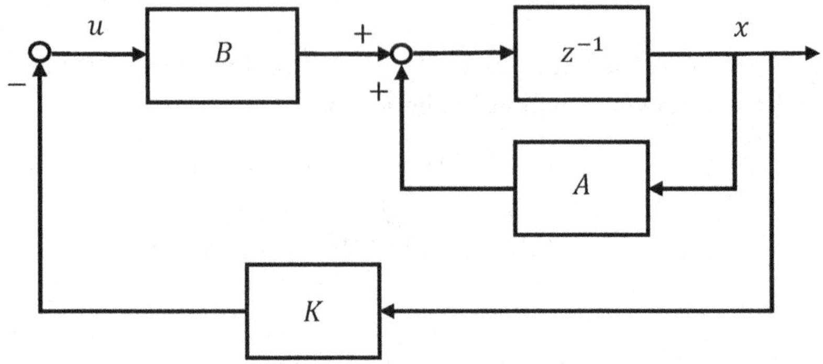

Figure 11.1: *LQ* control scheme.

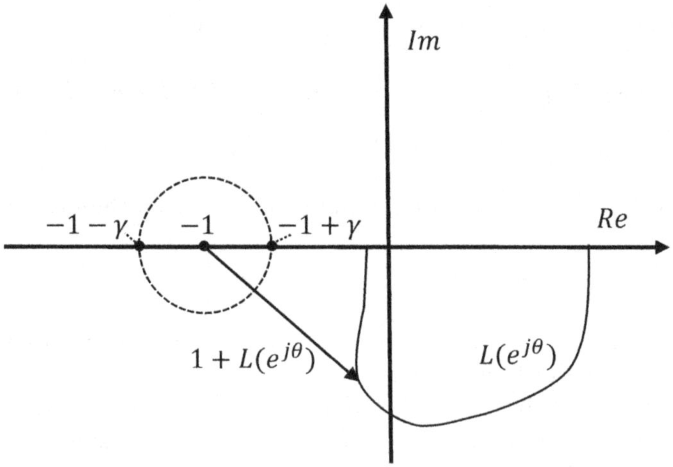

Figure 11.2: *LQ* control robustness.

11.3 Kalman predictor and filter

For discrete time systems it is possible to develop the Kalman theory to obtain a *predictor*, when the estimation of the state $x(k)$ depends on the output y and the input u up to the time instant $k-1$, or the *filter*, when the estimation of $x(k)$ depends on y and u up to the same time instant k.

It is assumed that the considered system is described by

$$\begin{aligned} x(k+1) &= Ax(k) + Bu(k) + v_x(k) \\ y(k) &= Cx(k) + v_y(k) \end{aligned}$$

where $v = \begin{bmatrix} v_x \\ v_y \end{bmatrix}$ is a white gaussian noise with null expected value and covariance matrix V, that is

$$E\left[v(k_1)v'(k_2)\right] = V\delta(k_1 - k_2)$$

where

$$V = \begin{bmatrix} \tilde{Q} & 0 \\ 0 & \tilde{R} \end{bmatrix}, \quad \tilde{Q} \in R^{n,n}, \quad \tilde{R} \in R^{p,p}, \quad \tilde{Q} \geq 0, \quad \tilde{R} > 0$$

Suppose also that the initial state $x(0)$ is a gaussian random variable with expected value $\bar{x}_0$ and covariance matrix

$$E\left[(x(0) - \bar{x}_0)(x(0) - \bar{x}_0)'\right] = \tilde{P}_0 \geq 0$$

and that $x(0)$, $v_x(k)$ e $v_y(k)$ are uncorrelated, that is

$$E\left[x(0)v'(k)\right] = 0, \quad \forall k \geq 0$$

11.3.1 Kalman predictor

Define by $\hat{x}(k|k-1)$ the one step prediction of $x(k)$, that is the estimation of the state x at time k carried out starting from the knowledge of the input u and output y until the time instant $k-1$.

In this case the predictor is given by

$$\hat{x}(k+1|k) = A\hat{x}(k|k-1) + Bu(k) + L(k)\left[y(k) - C\hat{x}(k|k-1)\right]$$

Let $\hat{e}(k|k-1) = x(k) - \hat{x}(k|k-1)$ be the estimation error; from the previous equations it follows

$$\begin{aligned} \hat{e}(k+1|k) &= \left[A - L(k)C\right]\hat{e}(k|k-1) + v_x(k) - L(k)v_y(k) \\ &= \left[A - L(k)C\right]\hat{e}(k|k-1) + \begin{bmatrix} I & -L(k) \end{bmatrix} v(k) \\ &= A_c(k)\hat{e}(k|k-1) + B_c(k)v(k) \end{aligned}$$

where

$$A_c(k) = \left[A - L(k)C\right], \quad B_c(k) = \begin{bmatrix} I & -L(k) \end{bmatrix}$$

from which

$$E\left[\hat{e}(k+1|k)\right] = A_c(k)E\left[\hat{e}(k|k-1)\right]$$

If the initial state of the predictor is chosen equal to

$$\hat{x}(0|-1) = \bar{x}_0$$

it results

$$E\left[\hat{e}(0|-1)\right] = 0$$

and

$$E\left[\hat{e}(k+1|k)\right] = 0, \quad \forall k$$

since the expected value of the estimation error is the state of an unforced system with null initial condition.

Now define the covariance matrix of the estimation error

$$\tilde{P}(k|k-1) = E\left[\hat{e}(k|k-1)\hat{e}'(k|k-1)\right]$$

with

$$\tilde{P}(0|-1) = \tilde{P}_0$$

The predictor gain $L(k)$ is chosen to minimize a generic quadratic form which depends on the covariance matrix of the error, that is to solve the following problem

$$\min_{L(k)} \gamma' \tilde{P}(k+1|k)\gamma$$

where γ is a generic vector. The following result holds.

Theorem 11.3.1. *The gain $L(k)$ minimizing $\gamma'\tilde{P}(k+1|k)\gamma$ is*

$$L(k) = A\tilde{P}(k|k-1)C'\left[C\tilde{P}(k|k-1)C' + \tilde{R}\right]^{-1}$$

where $\tilde{P}(k|k-1)$ is the solution to the Riccati equation

$$\begin{aligned} & \tilde{P}(k+1|k) \\ = \; & A\tilde{P}(k|k-1)A' + \tilde{Q} \\ - \; & A\tilde{P}(k|k-1)C'\left[C\tilde{P}(k|k-1)C' + \tilde{R}\right]^{-1} C\tilde{P}(k|k-1)A' \end{aligned}$$

with initial condition

$$\tilde{P}(0|-1) = \tilde{P}_0$$

Proof. From the error dynamic equation previously obtained it follows

$$\begin{aligned} & \tilde{P}(k+1|k) = E\left[\hat{e}(k+1|k)\hat{e}(k+1|k)'\right] \\ = \; & E\left[\left(A_c(k)\hat{e}(k|k-1) + B_c(k)v(k)\right)\left(A_c(k)\hat{e}(k|k-1) + B_c(k)v(k)\right)'\right] \\ = \; & A_c(k)E\left[\hat{e}(k|k-1)\hat{e}'(k|k-1)\right]A_c'(k) \\ & + B_c(k)E\left[v(k)v'(k)\right]B_c'(k) \\ & + 2A_c(k)E\left[\hat{e}(k|k-1)v'(k)\right]B_c'(k) \end{aligned}$$

In view of the hypothesis previously introduced, $\hat{e}(k|k-1) = x(k) - \hat{x}(k|k-1)$ is independent from $v(k)$; in fact $v(k)$ is a white noise and $\hat{e}(k|k-1)$ depends at most on $v(k-1)$. Hence, since $v(k)$ has null expected value, the last term of the previous expression is null. Moreover

$$\begin{aligned} & B_c(k)E[v(k)v'(k)]B_c'(k) \\ = & \begin{bmatrix} I & -L(k) \end{bmatrix} \begin{bmatrix} \tilde{Q} & 0 \\ 0 & \tilde{R} \end{bmatrix} \begin{bmatrix} I \\ -L'(k) \end{bmatrix} \\ = & \begin{bmatrix} \tilde{Q} & -L(k)\tilde{R} \end{bmatrix} \begin{bmatrix} I \\ -L'(k) \end{bmatrix} = \tilde{Q} + L(k)\tilde{R}L'(k) \end{aligned}$$

and

$$\begin{aligned} & \tilde{P}(k+1|k) \\ = & (A - L(k)C)\,\tilde{P}(k|k-1)\,(A - L(k)C)' + \tilde{Q} + L(k)\tilde{R}L'(k) \\ = & A\tilde{P}(k|k-1)A' - 2A\tilde{P}(k|k-1)C'L'(k) + L(k)C\tilde{P}(k|k-1)C'L'(k) \\ & + \tilde{Q} + L(k)\tilde{R}L'(k) \\ = & A\tilde{P}(k|k-1)A' + \tilde{Q} - 2A\tilde{P}(k|k-1)C'L'(k) \\ & + L(k)\left[C\tilde{P}(k|k-1)C' + \tilde{R}\right]L'(k) \end{aligned}$$

Thus, it is possible to derive with respect to $L(k)$ the quadratic form $\gamma'\tilde{P}(k+1|k)\gamma$ and to set to zero this derivative, so obtaining

$$-2\gamma' C\tilde{P}(k|k-1)A'\gamma + 2\gamma'\left[C\tilde{P}(k|k-1)C' + \tilde{R}\right]L'(k)\gamma = 0$$

from which

$$L(k) = A\tilde{P}(k|k-1)C'\left[C\tilde{P}(k|k-1)C' + \tilde{R}\right]^{-1}$$

Substituting this expression in the updating equation of $\tilde{P}(k+1|k)$ one obtains

$$\begin{aligned} & \tilde{P}(k+1|k) \\ = & A\tilde{P}(k|k-1)A' + \tilde{Q} \\ & - A\tilde{P}(k|k-1)C'\left[C\tilde{P}(k|k-1)C' + \tilde{R}\right]^{-1} C\tilde{P}(k|k-1)A' \end{aligned}$$

Now compare the Riccati equation for the predictor and the expression of the gain $L(k)$ with the results associated with the *LQ* control. By means of the following substitutions

LQ	KF
k	$-k$
A	A'
B	C'
Q	$\tilde{Q}$
R	$\tilde{R}$
P	$\tilde{P}$

and by duality considerations, the following result holds.

Theorem 11.3.2. *If*

1. *the pair* (A, B_q)*, with* B_q *such that* $\tilde{Q} = B_q B_q'$*, is reachable;*
2. *the pair* (A, C) *is observable;*

then

A) *the optimal predictor is*

$$\begin{aligned} \hat{x}(k+1|k) &= A\hat{x}(k|k-1) + Bu(k) + \bar{L}[y(k) - C\hat{x}(k|k-1)] \\ &= (A - \bar{L}C)\hat{x}(k|k-1) + Bu(k) + \bar{L}y(k) \end{aligned}$$

with

$$\bar{L} = A\bar{\tilde{P}}C' \left[C\bar{\tilde{P}}C' + \tilde{R}\right]^{-1}$$

and $\bar{\tilde{P}}$ *is the unique positive definite solution of the stationary Riccati equation*

$$\bar{\tilde{P}} = A\bar{\tilde{P}}A' + \tilde{Q} - A\bar{\tilde{P}}C' \left[C\bar{\tilde{P}}C' + \tilde{R}\right]^{-1} C\bar{\tilde{P}}A'$$

B) *the estimator is asymptotically stable, that is all the eigenvalues of* $(A - \bar{L}C)$ *have modulus less than 1.*

11.3.2 Kalman filter

With considerations similar to ones previously developed, it is possible to derive also the optimal filter $\hat{x}(k|k)$, that is the estimation of $x(k)$ given the inputs u and the outputs y up to the previous time instant k. Without going into technical details, in the following the main results are reported.

The filter is described by the following equations

$$\hat{x}(k+1|k+1) = A\hat{x}(k|k) + Bu(k) + L(k+1)\left[y(k+1) - C\left(A\hat{x}(k|k) + Bu(k)\right)\right]$$

where

$$\begin{aligned} L(k) &= \tilde{P}(k|k-1)C' \left[C\tilde{P}(k|k-1)C' + \tilde{R}\right]^{-1} \\ \tilde{P}(k|k-1) &= A\hat{P}(k-1|k-1)A' + \tilde{Q} \\ \hat{P}(k|k) &= \tilde{P}(k|k-1) - L(k)C\tilde{P}(k|k-1) \\ \hat{P}(0|0) &= \tilde{P}_0 \end{aligned}$$

and $\hat{P}(k|k)$ is the covariance of the estimation error $\hat{e}(k|k) = x(k) - \hat{x}(k|k)$.

Convergence and stability results analogous to the ones introduced for the predictor also hold for the filter.

11.3.3 Extended Kalman predictor

The approach adopted to derive the Kalman predictor for linear systems can be extended also to nonlinear systems affected by stochastic disturbances. In this way it is possible to obtain the extended Kalman predictor (filter), widely used and known with the acronym *EKF* (Extended Kalman Filter), that will be presented in the following.

Suppose that the considered system is described by the nonlinear equations

$$\begin{aligned} x(k+1) &= f(x(k), u(k)) + v_x(k) \\ y(k) &= g(x(k)) + v_y(k) \end{aligned}$$

where the functions f and g are continuously differentiable with respect to their arguments, while, as before, $v = \begin{bmatrix} v_x \\ v_y \end{bmatrix}$ is a white gaussian noise with null expected value and covariance matrix V, that is

$$E\left[v(k_1)v'(k_2)\right] = V\delta(k_1 - k_2)$$

and

$$V = \begin{bmatrix} \tilde{Q} & 0 \\ 0 & \tilde{R} \end{bmatrix}, \quad \tilde{Q} \in R^{n,n}, \quad \tilde{R} \in R^{p,p}, \quad \tilde{Q} \geq 0, \quad \tilde{R} > 0$$

Suppose, also in this case, that the initial state $x(0)$ is a gaussian random variable with expected value $\bar{x}_0$ and covariance matrix

$$E\left[(x(0) - \bar{x}_0)(x(0) - \bar{x}_0)'\right] = \tilde{P}_0 \geq 0$$

and that $x(0)$, $v_x(k)$ and $v_y(k)$ are uncorrelated, that is

$$E\left[x(0)v'(k)\right] = 0, \quad \forall k \geq 0$$

The goal is to design a predictor made by a term defined by the nonlinear system dynamics, plus a correction given by the difference between the measured output and the estimated one, multiplied by a suitable gain.

Linearized predictor

First a solution that, even if not very effective, is useful to present the general concepts is introduced. Later the structure of the standard *EKF* will be discussed.

Define by $\bar{x}(k)$ and $\bar{y}(k)$ the solution to the state and output equations with null noises, that is $v(k) = 0$, $\forall k \geq 0$. Therefore

$$\begin{aligned} \bar{x}(k+1) &= f(\bar{x}(k), u(k)) \\ \bar{y}(k) &= g(\bar{x}(k)) \end{aligned}$$

Letting $\delta x(k) = x(k) - \bar{x}(k)$, $\delta y(k) = y(k) - \bar{y}(k)$ and linearizing with respect to $\bar{x}(k)$ and $\bar{y}(k)$ leads to

$$\begin{aligned}\bar{x}(k+1) + \delta x(k+1) &\simeq f(\bar{x}(k), u(k)) + \left.\frac{\partial f(x,u)}{\partial x}\right|_{\bar{x}(k),u(k)} \delta x(k) + v_x(k) \\ \bar{y}(k) + \delta y(k) &\simeq g(\bar{x}(k)) + \left.\frac{\partial g(x)}{\partial x}\right|_{\bar{x}(k)} \delta x(k) + v_y(k)\end{aligned}$$

Recalling the previous relations and defining

$$\bar{A}(k) = \left.\frac{\partial f(x,u)}{\partial x}\right|_{\bar{x}(k),u(k)}, \quad \bar{C}(k) = \left.\frac{\partial g(x)}{\partial x}\right|_{\bar{x}(k)}$$

the linearized system can be rewritten in the form

$$\begin{aligned}\delta x(k+1) &= \bar{A}(k)\delta x(k) + v_x(k) \\ \delta y(k) &= \bar{C}(k)\delta x(k) + v_y(k)\end{aligned}$$

For this linearized system it is possible to use the linear Kalman predictor

$$\delta\hat{x}(k+1|k) = \bar{A}(k)\delta\hat{x}(k|k-1) + \bar{L}(k)\left[\delta y(k) - \bar{C}(k)\delta\hat{x}(k|k-1)\right]$$

where the gain $\bar{L}(k)$ is computed with the formula of the linear Kalman predictor applied to time-variant systems, that is

$$\bar{L}(k) = \bar{A}(k)\tilde{P}(k|k-1)\bar{C}(k)'\left[\bar{C}(k)\tilde{P}(k|k-1)\bar{C}(k)' + \tilde{R}\right]^{-1}$$

where $\tilde{P}(k|k-1)$ is the solution to the Riccati equation

$$\begin{aligned}&\tilde{P}(k+1|k) \\ = \;& \bar{A}(k)\tilde{P}(k|k-1)\bar{A}(k)' + \tilde{Q} \\ - \;& \bar{A}(k)\tilde{P}(k|k-1)\bar{C}(k)'\left[\bar{C}(k)\tilde{P}(k|k-1)\bar{C}(k)' + \tilde{R}\right]^{-1}\bar{C}(k)\tilde{P}(k|k-1)\bar{A}(k)'\end{aligned}$$

with initial condition

$$\tilde{P}(0|-1) = \tilde{P}_0$$

Then, given the estimate $\delta\hat{x}(k|k-1)$, at any instant it is possible to set

$$\hat{x}(k|k-1) = \bar{x}(k) + \delta\hat{x}(k|k-1)$$

and, recalling that $\delta y(k) = y(k) - \bar{y}(k) = y(k) - g(\bar{x}(k))$, it follows

$$\begin{aligned}\hat{x}(k+1|k) &= f(\bar{x}(k), u(k)) + \bar{A}(k)\delta\hat{x}(k|k-1) \\ &+ \bar{L}(k)\left[y(k) - g(\bar{x}(k)) - \bar{C}(k)\delta\hat{x}(k|k-1)\right]\end{aligned}$$

Since

$$f(\hat{x}(k|k-1), u(k)) \simeq f(\bar{x}(k), u(k)) + \bar{A}(k)\delta\hat{x}(k|k-1)$$

and

$$g(\hat{x}(k|k-1)) \simeq g(\bar{x}(k)) + \bar{C}(k)\delta\hat{x}(k|k-1)$$

the final expression of the predictor becomes

$$\hat{x}(k+1|k) = f\left(\hat{x}(k|k-1), u(k)\right) + \bar{L}(k)\left[y(k) - g(\hat{x}(k|k-1))\right]$$

that represents the linearized Kalman predictor.

As anticipated, the predictor is characterized by the nonlinear dynamics of the system, plus a corrective term given by the output error, also computed with the nonlinear output transformation, multiplied by the time varying gain $\bar{L}(k)$. Only this term is computed based on the linearized model computed at any time instant with the nominal solution to the state equation (matrices $\bar{A}(k)$ and $\bar{C}(k)$).

The linearized predictor often does not guarantee satisfactory performances because the computation of the matrices $\bar{A}(k)$ and $\bar{C}(k)$ is based on the nominal solution to the state equation. In this way, the available measures are not used and, due to the presence of noise, $\bar{x}(k)$ can be very different from $x(k)$, so that the linearized model can be very poor.

Extended Kalman predictor

To overcome the problems associated with the linearized predictor, usually the following predictor is used

$$\hat{x}(k+1|k) = f\left(\hat{x}(k|k-1), u(k)\right) + \hat{L}(k)\left[y(k) - g(\hat{x}(k|k-1))\right]$$

where the gain $\hat{L}(k)$ is computed using the matrices of the system linearized around the most recent available state estimation. Hence the matrices

$$\hat{A}(k) = \left.\frac{\partial f(x,u)}{\partial x}\right|_{\hat{x}(k|k-1),u(k)}, \quad \hat{C}(k) = \left.\frac{\partial g(x)}{\partial x}\right|_{\hat{x}(k|k-1)}$$

are used in the update formula

$$\hat{L}(k) = \hat{A}(k)\tilde{P}(k|k-1)\hat{C}(k)'\left[\hat{C}(k)\tilde{P}(k|k-1)\hat{C}(k)' + \tilde{R}\right]^{-1}$$

where $\tilde{P}(k|k-1)$ is the unique definite positive solution to the Riccati equation $\tilde{P}(0|-1) = \tilde{P}_0$ and

$$\begin{aligned} & \tilde{P}(k+1|k) \\ = \; & \hat{A}(k)\tilde{P}(k|k-1)\hat{A}(k)' + \tilde{Q} + \\ - \; & \hat{A}(k)\tilde{P}(k|k-1)\hat{C}(k)'\left[\hat{C}(k)\tilde{P}(k|k-1)\hat{C}(k)' + \tilde{R}\right]^{-1}\hat{C}(k)\tilde{P}(k|k-1)\hat{A}(k)' \end{aligned}$$

In this way, the gain $\hat{L}(k)$ depends on the current estimation $\hat{x}(k|k-1)$ through the matrices $\hat{A}(k)$ and $\hat{C}(k)$.

Estimation of unknown parameters with the *EKF*

The extended Kalman predictor can also be used to estimate unknown parameters with constant value. Consider for example the system

$$\begin{aligned} x(k+1) &= f\left(x(k), u(k), \alpha\right) + v_x(k) \\ y(k) &= g\left(x(k), \alpha\right) + v_y(k) \end{aligned}$$

where α is an unknown constant parameter. To jointly estimate the state of the system and the value of α the fictitious dynamic

$$\alpha(k+1) = \alpha(k)$$

is introduced and the following enlarged state

$$\tilde{x} = \begin{bmatrix} x \\ \alpha \end{bmatrix}$$

is defined. The enlarged system is given by

$$\begin{aligned} \tilde{x}(k+1) &= \tilde{f}\left(\tilde{x}(k), u(k)\right) + \tilde{v}_x(k) \\ y(k) &= \tilde{g}\left(\tilde{x}(k)\right) + v_y(k) \end{aligned}$$

where $\tilde{f}$, $\tilde{g}$, and $\tilde{v}_x$ can be easily obtained from the previous relations.

It is possible to apply to this system the *EKF* theory to obtain an estimation of $\tilde{x}$, that is of x and α.

11.4 *LQG* control

Consider the system

$$\begin{aligned} x(k+1) &= Ax(k) + Bu(k) + v_x(k) \\ y(k) &= Cx(k) + v_y(k) \end{aligned}$$

where the noise $v = \begin{bmatrix} v_x \\ v_y \end{bmatrix}$ and the initial state $x(0)$ satisfy the assumption adopted to derive the Kalman predictor and filter.

For this system, the following cost function

$$J = \lim_{\bar{k}\to\infty} \frac{1}{\bar{k}} E\left[\sum_{k=0}^{\bar{k}-1} x'(k)Qx(k) + u'(k)Ru(k)\right]$$

must be minimized when the state x is not measurable.

Under the assumption that all the conditions that guarantee the stability of both *LQ* and *KF*, it is possible to use a "separation principle" that consists of the following steps:

1. determine with the stationary Kalman predictor (or filter) the Kalman gain $\bar{L}$ and the estimate $\hat{x}(k|k-1)$ ($\tilde{x}(k|k)$);

2. compute the optimal LQ control law, that is the gain $\bar{K}$;

3. apply the control law

$$u(k) = -\bar{K}\hat{x}(k\,|k-1) \text{ in the predictor case}$$

or

$$u(k) = -\bar{K}\tilde{x}(k\,|k) \text{ in the filter case.}$$

It is possible to show that the resulting closed-loop system is asymptotically stable with the eigenvalues of $(A - B\bar{K})$ and, if the predictor is considered, the ones of $(A - \bar{L}C)$.

12
Model Predictive Control

In this chapter the basic principles of Model Predictive Control, or *MPC*, are presented. *MPC* is a family of algorithms that has had an enormous industrial impact in the last thirty years. Today, several commercial *MPC* control algorithms are available, and *MPC* is the most widely used advanced control methodology in the process industry.

The most important characteristics of *MPC*, which make it very flexible and suitable for a large number of applications, are the possibility to:

- formulate the control problem as an optimization one, where different goals can be considered;
- explicitly include in the control problem formulation state and input constraints;
- synthesize the controller based on process models empirically derived by means of simple experiments, such as step or impulse responses.

The main ingredients of an *MPC* algorithm are:

- a process model, usually in discrete-time;
- input, output, and state constraints;
- a cost function J defined, at any time instant k, over a finite horizon $[k, k+N]$;
- an optimization algorithm computing the future optimal control sequence;
- the so-called *Receding Horizon* (RH) principle: *at any time instant* k*, based on the available process information, solve the optimization problem with respect to the future control sequence* $[u(k), ..., u(k+N-1)]$ *and apply only its first element* $u^o(k)$*. Then, at the next time instant* $k+1$*, a new optimization problem is solved, based on the process information available at time* $k+1$*, along the prediction horizon* $[k+1, k+N]$. Note that, by means of this strategy, a time invariant feedback control law is obtained even if a finite horizon optimization problem is solved at any time instant.

In the following, some general concepts and basic *MPC* algorithms are introduced following the ideas developed in the previous Chapters for linear optimal control. Then a number of extensions and industrial approaches are described. Finally, the development of *MPC* algorithms with guaranteed stabilizing properties are briefly discussed.

12.1 *MPC* of linear systems in state space form

Consider the system

$$\begin{aligned} x(k+1) &= Ax(k) + Bu(k) \\ y(k) &= Cx(k) \end{aligned}$$

where the state $x \in R^n$ is assumed to be measurable, $u \in R^m$ is the control variable, and $y \in R^p$ is the output variable.

The optimization problem consists of computing, at each time instant k, the control sequence $u(k)$, $u(k+1)$, ..., $u(k+N-1)$ that minimizes the finite horizon quadratic cost function

$$J(x(k), u(\cdot), k) = \sum_{i=0}^{N-1} \left(\|x(k+i)\|_Q^2 + \|u(k+i)\|_R^2 \right) + \|x(k+N)\|_S^2 \qquad (12.1)$$

where $Q = Q' \geq 0, R = R' > 0,\ S = S' \geq 0$ are matrices of suitable dimensions. In *MPC*, the positive integer N is usually called *prediction horizon*.

12.1.1 Closed-loop solution

As discussed in Chapter 11 with reference to LQ control, the solution of the stated (unconstrained) problem is given by the time-varying optimal control law

$$u^o(k+i) = -K(i)x(k+i), \quad i = 0, 1, ..., N-1 \qquad (12.2)$$

where

$$K(i) = (R + B'P(i+1)B)^{-1} B'P(i+1)A$$

and $P(i)$ is the solution to the difference Riccati equation

$$P(i) = Q + A'P(i+1)A - A'P(i+1)B\,(R + B'P(i+1)B)^{-1} B'P(i+1)A$$

with boundary condition

$$P(N) = S$$

Equation (12.2) is a *closed-loop* control law, since the control variable $u(k+i)$ depends on the state at time $x(k+i)$. In view of the above stated RH principle, only the first element of the optimal future control sequence is applied, so that the *MPC* control law is state-feedback, time-invariant, and given by

$$u^{MPC}(k) = -K(0)x(k)$$

12.1.2 Open-loop solution

The solution to the optimization problem (12.1) can also be derived recalling that, in view of Lagrange equation,

$$x(k+i) = A^i x(k) + \sum_{j=0}^{i-1} A^{i-j-1} B u(k+j), \quad i > 0$$

Letting

$$X(k) = \begin{bmatrix} x(k+1) \\ x(k+2) \\ \vdots \\ x(k+N-1) \\ x(k+N) \end{bmatrix}, \mathcal{A} = \begin{bmatrix} A \\ A^2 \\ \vdots \\ A^{N-1} \\ A^N \end{bmatrix}, U(k) = \begin{bmatrix} u(k) \\ u(k+1) \\ \vdots \\ u(k+N-2) \\ u(k+N-1) \end{bmatrix}$$

$$\mathcal{B} = \begin{bmatrix} B & 0 & 0 & \cdots & 0 & 0 \\ AB & B & 0 & \cdots & 0 & 0 \\ \cdots & \cdots & \cdots & \cdots & \cdots & \cdots \\ A^{N-2}B & A^{N-3}B & A^{N-4}B & \cdots & B & 0 \\ A^{N-1}B & A^{N-2}B & A^{N-3}B & \cdots & AB & B \end{bmatrix}$$

it follows that

$$X(k) = \mathcal{A}x(k) + \mathcal{B}U(k) \tag{12.3}$$

Moreover, define the matrices, with N blocks on the diagonal,

$$\mathcal{Q} = \begin{bmatrix} Q & 0 & \cdots & 0 & 0 \\ 0 & Q & \cdots & 0 & 0 \\ \cdots & \cdots & \ddots & \cdots & \cdots \\ 0 & 0 & \cdots & Q & 0 \\ 0 & 0 & \cdots & 0 & S \end{bmatrix}, \quad \mathcal{R} = \begin{bmatrix} R & 0 & \cdots & 0 & 0 \\ 0 & R & \cdots & 0 & 0 \\ \cdots & \cdots & \ddots & \cdots & \cdots \\ 0 & 0 & \cdots & R & 0 \\ 0 & 0 & \cdots & 0 & R \end{bmatrix} \tag{12.4}$$

and observe that

$$arg(\min_{u(\cdot)} J(x(k), u(\cdot), k)) = arg(\min_{U(k)} \bar{J}(x(k), U(k), k))$$

where

$$\bar{J}(x(k), U(k), k) = X'(k)\mathcal{Q}X(k) + U'(k)\mathcal{R}U(k)$$

since the term $x'(k)Qx(k)$ in $J(x(k), u(\cdot), k)$, see (12.1), does not depend on $u(k+j)$, $j \geq 0$, and can be neglected.

Recalling (12.3), it follows that

$$\begin{aligned} \bar{J}(x(k), U(k), k) &= (\mathcal{A}x(k) + \mathcal{B}U(k))' \, \mathcal{Q} \, (\mathcal{A}x(k) + \mathcal{B}U(k)) + U'(k)\mathcal{R}U(k) \\ &= x'(k)\mathcal{A}'\mathcal{Q}\mathcal{A}x(k) + 2x'(k)\mathcal{A}'\mathcal{Q}\mathcal{B}U(k) + U'(k)\left(\mathcal{B}'\mathcal{Q}\mathcal{B} + \mathcal{R}\right)U(k) \end{aligned}$$

This is a positive definite quadratic form, since $\mathcal{R} > 0$, and its minimum can be computed by setting to zero its derivative with respect to $U(k)$. The resulting optimal control sequence $U^o(k)$ is

$$U^o(k) = -\left(\mathcal{B}'\mathcal{Q}\mathcal{B} + \mathcal{R}\right)^{-1}\mathcal{B}'\mathcal{Q}\mathcal{A}x(k)$$

Note that this solution depends on the prediction of the state, along the considered horizon, based on the current state $x(k)$. In fact, defining

$$\mathcal{K} = \left(\mathcal{B}'\mathcal{Q}\mathcal{B} + \mathcal{R}\right)^{-1}\mathcal{B}'\mathcal{Q}\mathcal{A} = \begin{bmatrix} \mathcal{K}(0) \\ \mathcal{K}(1) \\ \vdots \\ \mathcal{K}(N-1) \end{bmatrix}, \quad \mathcal{K}(i) \in R^{m,n}$$

it is possible to write

$$U^o(k) = -\begin{bmatrix} \mathcal{K}(0) \\ \mathcal{K}(1) \\ \vdots \\ \mathcal{K}(N-1) \end{bmatrix} x(k)$$

or

$$u^o(k+i) = -\mathcal{K}(i)x(k), \quad i = 0, 1, ..., N-1 \tag{12.5}$$

For this reason the control law (12.5) results in an *open-loop* solution, at least for $i > 0$. However, by applying the RH principle, also in this case a state-feedback control law is obtained

$$u^{MPC}(k) = -\mathcal{K}(0)x(k)$$

Note that:

- in the nominal case here considered, that is in absence of disturbances or modeling errors, the closed-loop solution (12.2) and the open-loop one (12.5) coincide, while in the presence of disturbances or model uncertainty only the first element of (12.2) and (12.5) are guaranteed to be equal, that is $\mathcal{K}(0) = K(0)$;

- the open-loop solution (12.5) has been derived by computing the prediction of the future states based on the value of the current state $x(k)$. This structure can be easily generalized to compute the model predictive control law starting from different model representations, like the transfer function, the impulse or step response.

12.1.3 Handling state, control, and output constraints

The main difference between the open- and closed-loop approaches is that in the first one the presence of state, input and output constraints can be easily included into the optimization problem. In the following it will be assumed that the future inputs, states, and/or outputs must satisfy the following constraints

$$u_m \leq u(k+i) \leq u_M, \quad i = 0, ..., N-1$$

$$x_m \leq x(k+i) \leq x_M, \quad i = 1, ..., N$$

$$y_m \leq y(k+i) \leq y_M, \quad i = 1, ..., N$$

where the vector inequalities must be considered element by element. Defining the following vectors, made by the concatenation of N vector elements,

$$U_m = \begin{bmatrix} u_m \\ u_m \\ \vdots \\ u_m \end{bmatrix}, \quad U_M = \begin{bmatrix} u_M \\ u_M \\ \vdots \\ u_M \end{bmatrix}, \quad X_m = \begin{bmatrix} x_m \\ x_m \\ \vdots \\ x_m \end{bmatrix}, \quad X_M = \begin{bmatrix} x_M \\ x_M \\ \vdots \\ x_M \end{bmatrix}$$

$$Y_m = \begin{bmatrix} y_m \\ y_m \\ \vdots \\ y_m \end{bmatrix}, \quad Y_M = \begin{bmatrix} y_M \\ y_M \\ \vdots \\ y_M \end{bmatrix}, \quad Y(k) = \begin{bmatrix} y(k+1) \\ y(k+2) \\ \vdots \\ y(k+N-1) \\ y(k+N) \end{bmatrix}$$

the finite horizon optimization problem can be reformulated as

$$\min_{U(k)} \quad \bar{J}(x(k), U(k), k) = (\mathcal{A}x(k) + \mathcal{B}U(k))' \, \mathcal{Q} \, (\mathcal{A}x(k) + \mathcal{B}U(k)) + U'(k)\mathcal{R}U(k)$$

with constraints

$$\begin{aligned} X_m &\leq X(k) = \mathcal{A}x(k) + \mathcal{B}U(k) \leq X_M \\ U_m &\leq U(k) \leq U_M \\ Y_m &\leq Y(k) \leq Y_M \end{aligned}$$

where $Y(k)$ is immediately obtained from $X(k)$ by recalling the output transformation. This constrained optimization problem does not admit an explicit solution, but it can be easily solved by means of quadratic programming methods to obtain the optimal vector $U^o(k)$ of the future values of the control variable along the prediction horizon.

12.2 Industrial *MPC* algorithms

The basic *MPC* formulation previously introduced is now extended to cope with tracking problems and systems affected by disturbances. Moreover, a number of empirical rules typical of industrial applications are described.

12.2.1 Tracking of reference signals and disturbances

Let the system be described by

$$\begin{aligned} x(k+1) &= Ax(k) + Bu(k) + Md(k) \\ y(k) &= Cx(k) + d(k) \end{aligned}$$

where $d \in R^p$ is a disturbance, assumed to be known at the current time instant k and, possibly, in the future prediction horizon.

The goal is to track with the output variable a reference signal y^o in the future prediction horizon. To this end, note that

$$x(k+i) = A^i x(k) + \sum_{j=0}^{i-1} A^{i-j-1} \left(Bu(k+j) + Md(k+j) \right)$$

and

$$y(k+i) = Cx(k+i) + d(k+i)$$

The quadratic cost function to be minimized must include a term weighting the future error variable, i.e.

$$\begin{aligned} J(x(k), u(\cdot), k) &= \sum_{i=0}^{N-1} \left(\|y^o(k+i) - y(k+i)\|_Q^2 + \|u(k+i)\|_R^2 \right) \\ &\quad + \|y^o(k+N) - y(k+N)\|_S^2 \end{aligned}$$

where $Q = Q' \geq 0$, $R = R' > 0$, $S = S' \geq 0$ are matrices of suitable dimensions.

Based on the previous formula, it is possible to write

$$Y(k) = \mathcal{A}_c x(k) + \mathcal{B}_c U(k) + \mathcal{M}_c D(k)$$

where

$$\mathcal{A}_c = \begin{bmatrix} CA \\ CA^2 \\ \vdots \\ CA^{N-1} \\ CA^N \end{bmatrix}, \quad D(k) = \begin{bmatrix} d(k) \\ d(k+1) \\ \vdots \\ d(k+N-2) \\ d(k+N-1) \\ d(k+N) \end{bmatrix}$$

$$\mathcal{B}_c = \begin{bmatrix} CB & 0 & 0 & \cdots & 0 & 0 \\ CAB & CB & 0 & \cdots & 0 & 0 \\ \cdots & \cdots & \cdots & \cdots & \cdots & \cdots \\ CA^{N-2}B & CA^{N-3}B & CA^{N-4}B & \cdots & CB & 0 \\ CA^{N-1}B & CA^{N-2}B & CA^{N-3}B & \cdots & CAB & CB \end{bmatrix}$$

$$\mathcal{M}_c = \begin{bmatrix} CM & I & 0 & \cdots & 0 & 0 & 0 \\ CAM & CM & I & \cdots & 0 & 0 & 0 \\ \cdots & \cdots & \cdots & \cdots & \cdots & \cdots & \cdots \\ CA^{N-2}M & CA^{N-3}M & CA^{N-4}M & \cdots & CM & I & 0 \\ CA^{N-1}M & CA^{N-2}M & CA^{N-3}M & \cdots & CAM & CM & I \end{bmatrix}$$

Now, letting

$$Y^o(k) = \begin{bmatrix} y^o(k+1) \\ y^o(k+2) \\ \vdots \\ y^o(k+N-1) \\ y^o(k+N) \end{bmatrix}$$

and defining the matrices $\mathcal{Q}$ e $\mathcal{R}$ as in (12.4), one obtains

$$arg(\min_{U(k)} J(x(k), u(\cdot), k)) = arg(\min_{U(k)} \bar{J}(x(k), U(k), k))$$

where

$$\bar{J}(x(k), U(k), k) = (Y^o(k) - Y(k))' \, \mathcal{Q} \, (Y^o(k) - Y(k)) + U'(k)\mathcal{R}U(k) \tag{12.6}$$

or

$$\begin{aligned} &\bar{J}(x(k), U(k), k) \\ &= (Y^o(k) - \mathcal{A}_c x(k) - \mathcal{B}_c U(k) - \mathcal{M}_c D(k))' \, \mathcal{Q} \, (Y^o(k) - \mathcal{A}_c x(k) - \mathcal{B}_c U(k) - \mathcal{M}_c D(k)) \\ &+ U'(k)\mathcal{R}U(k) \end{aligned}$$

This cost function must be minimized, as before, with respect to the future values $u(k+i)$, $i \geq 0$ of the control, and possibly under suitable constraints on x, u, y. Note however that it depends on the future reference signals and disturbances, which often are unknown. In these cases, it is customary to assume that

$$y^o(k+i) = y^o(k) \quad , \quad d(k+i) = d(k)$$

12.2.2 Constraints relaxation

When the performance index (12.6) must be minimized under output and control constraints such that

$$\begin{aligned} U_m &\leq U(k) \leq U_M \\ Y_m &\leq Y(k) \leq Y_M \end{aligned}$$

it can happen that a feasible solution does not exist due to the presence of disturbances. For this reason, it is more convenient to soften the output constraints by including in the optimization problem suitable slack variables. For simplicity, only one slack variable ε is used in the following, and the problem is stated as

$$\min_{U(k),\varepsilon} (Y^o(k) - Y(k))' \, \mathcal{Q} \, (Y^o(k) - Y(k)) + U'(k)\mathcal{R}U(k) + \rho\varepsilon$$

subject to

$$\begin{aligned} U_m &\leq U(k) \leq U_M \\ Y_m - \varepsilon \boldsymbol{I} &\leq Y(k) \leq Y_M + \varepsilon \boldsymbol{I} \\ \varepsilon &\geq 0 \end{aligned}$$

where $I = \begin{bmatrix} 1 & 1 & \dots & 1 \end{bmatrix}'$ and ρ must be selected sufficiently high to guarantee that the optimum is $\varepsilon = 0$ when a feasible solution exists.

12.2.3 Constant reference signals and integral action

Assume that the reference signal is constant, i.e. $y^o(k) = y^o$. In this case, provided that $p \leq m$ and the system does not have any invariant zero $z = 1$, it is possible to compute a steady-state pair $(\bar{x}, \bar{u})$ such that

$$\begin{aligned} \bar{x} &= A\bar{x} + B\bar{u} \\ y^o &= C\bar{x} \end{aligned}$$

and it is worth choosing the performance index to be minimized as

$$J(x(k), u(\cdot), k) = \sum_{i=0}^{N-1} \left(\|y^o - y(k+i)\|_Q^2 + \|u(k+i) - \bar{u}\|_R^2 \right) \tag{12.7}$$

$$+ \|y^o - y(k+N)\|_S^2 \tag{12.8}$$

so that $J = 0$ for $y(k+i) = y^o$ and $u(k+i) = \bar{u}$. It is then easy to extend the previous considerations to this case, and to reformulate the constrained optimization problem. However, the computation of the steady-state pair $(\bar{x}, \bar{u})$ leading to a zero output error cannot be performed in case of unknown disturbances and/or modeling errors. For this reason, it is advisable to resort to alternative formulations guaranteeing that the (implicit) *MPC* control law incorporates an effective integral action. To this end, the two main approaches described in the following have gained wide popularity.

Disturbance estimation

Let assume that the system is affected by an unknown constant disturbance d and described by

$$\begin{aligned} x(k+1) &= Ax(k) + Bu(k) + Md(k) \\ d(k+1) &= d(k) \\ y(k) &= Cx(k) + Fd(k) \end{aligned}$$

where different choices can be made for the matrices M and F to represent the dependence of the state and the output on the disturbance d. For example, the choice $M = B$ and $F = 0$ models input disturbances, while $M = 0$ and $F = I$ describes additive disturbances on the output.
As discussed in the previous chapters, the disturbance (and possibly the state) can be estimated with an observer provided that the enlarged system is observable. To this regard, first recall (see Section 6.2.2) that the dimension r of the disturbance must be not greater than the number of outputs p, i.e. $r \leq p$. Moreover, the enlarged system must be observable (or at least detectable); to this

regard the following result can be easily proven with the *PBH* test

Theorem 12.2.1. *The pair*

$$\bar{A} = \begin{bmatrix} A & M \\ 0 & I \end{bmatrix} \quad , \quad \bar{C} = \begin{bmatrix} C & F \end{bmatrix}$$

is observable if and only if the pair (A, C) *is observable and*

$$rank \begin{bmatrix} A - I & M \\ C & F \end{bmatrix} = n + r$$

Under the conditions of this theorem, it is possible to compute, with a proper observer and at any time instant k, the estimate $\begin{bmatrix} \hat{x}(k) \\ \hat{d}(k) \end{bmatrix}$, and then to obtain the state and input steady-state pair corresponding to the equilibrium specified by a future constant disturbance ($d(k+i) = \hat{d}(k)$, $i \geq 0$)

$$\begin{bmatrix} I - A & -B \\ C & 0 \end{bmatrix} \begin{bmatrix} \bar{x} \\ \bar{u} \end{bmatrix} = \begin{bmatrix} M & 0 \\ -F & I \end{bmatrix} \begin{bmatrix} \hat{d}(k) \\ y^o \end{bmatrix}$$

Finally, the pair $(\bar{x}, \bar{u})$ allows one to formulate the optimization problem (12.7).

Explicit integral action

A different approach, which resembles the solutions adopted in the previous chapters, is based on the scheme shown in Figure 12.1.

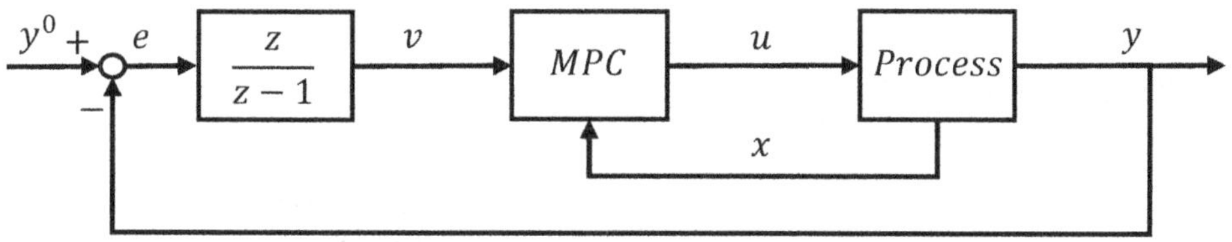

Figure 12.1: Control scheme with integral action

Consider the system enlarged with an integral action

$$\begin{aligned} x(k+1) &= Ax(k) + Bu(k) \\ v(k+1) &= v(k) + e(k+1) \end{aligned}$$

where $e(k) = y^o - y(k)$. Equivalently

$$\begin{aligned} x(k+1) &= Ax(k) + Bu(k) \\ v(k+1) &= v(k) + y^o - Cx(k+1) = v(k) + y^o - CAx(k) - CBu(k) \end{aligned}$$

By defining the state $\delta x(k) = x(k) - x(k-1)$ and control $\delta u(k) = u(k) - u(k-1)$ increments, by subtracting to the expressions above their value at the previous time instant, and by recalling that $v(k+1) - v(k) = e(k+1)$, the enlarged system can be written as

$$\begin{bmatrix} \delta x(k+1) \\ e(k+1) \end{bmatrix} = \begin{bmatrix} A & 0 \\ -CA & I \end{bmatrix} \begin{bmatrix} \delta x(k) \\ e(k) \end{bmatrix} + \begin{bmatrix} B \\ -CB \end{bmatrix} \delta u(k)$$

It is easy to prove that the pair $\begin{bmatrix} A & 0 \\ -CA & I \end{bmatrix}, \begin{bmatrix} B \\ -CB \end{bmatrix}$ is reachable provided that the pair (A, B) is reachable and the system under control does not possess invariant zeros at $z = 1$ (derivative actions).

For this enlarged system, usually called *in velocity form*, the following performance index can be considered

$$\begin{aligned} \min_{\delta u(k+i), i=0,\dots,N-1} J(\delta x(k), \delta u(\cdot), k) &= \sum_{i=0}^{N-1} \left(\|e(k+i)\|_Q^2 + \|\delta u(k+i)\|_R^2 \right) \\ &+ \|e(k+N)\|_S^2 \end{aligned}$$

subject to constraints on the future state and control variables and increments. This formulation guarantees that $J = 0$ when $e(k+i) = 0$, i.e. $y(k+i) = y^o$, and $\delta u(k+i) = 0$, i.e. u is constant.

Note that $x(k+i)$ and $u(k+i)$ can be easily written as functions of $x(k+i-1)$, $\delta x(k+i)$, $u(k+i-1)$, $\delta u(k+i)$. Finally, once the future optimal sequence $\delta u^o(k+i)$, $i = 0, \dots, N-1$ of control increments is computed, and according to the receding horizon approach, the control variable

$$u(k) = u(k-1) + \delta u^o(k)$$

is effectively used.

Some comments are in order:

- this approach does not require to estimate the disturbance d and/or to compute the corresponding steady state $(\bar{x}, \bar{u})$ at any new variation of the reference signal;
- constraints on δu can be very helpful to describe physical restrictions on the rate of variation of the control variables;
- additional terms weighting the future state increments can be easily included into the performance index to be minimized.

12.2.4 Control horizon

A large value of N is often chosen to include in the prediction horizon all the main process dynamics, so that also the number of (vector) optimization variables $u(k)$, $u(k+1)$, ..., $u(k+N-1)$ is large, at the price of an heavy computational load associated with the solution to the optimization problem to be solved on-line. For this reason, and in order to obtain a less aggressive control action, many industrial algorithms allow one to define a control horizon $0 < N_u < N$ and to consider the additional constraints

$$u(k+i) = u(k+i-1), \quad i = N_u, ..., N-1$$

or

$$\delta u(k+i) = 0, \quad i = N_u, ..., N-1$$

when the system is described in velocity form.
Correspondingly, the performance index to be considered, for example in the case of systems in velocity form, becomes

$$\begin{aligned} J(x(k), u(\cdot), k) &= \sum_{i=0}^{N-1} \|y^o(k+i) - y(k+i)\|_Q^2 \\ &+ \sum_{i=0}^{N_u-1} \|\delta u(k+i)\|_R^2 \\ &+ \|y^o(k+N) - y(k+N)\|_S^2 \end{aligned}$$

and all the previous developments can be easily extended to include in the problem formulation this new set of constraints. Note that these constraints can be explicitly considered when computing the output predictions.

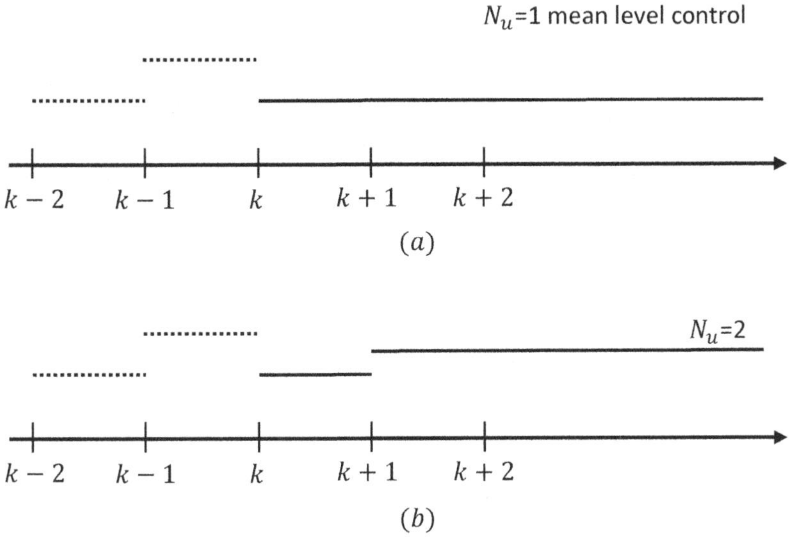

Figure 12.2: Control variable with $N_u = 1$ (a) and $N_u = 2$ (b).

A typical example is the so-called *Mean Level Control*, largely used and based on the choice $N_u = 1$. In this algorithm, only one control variation is allowed for, while it is assumed that the control variable will remain constant over the whole prediction horizon, see Figure 12.2. However, due to the Receding Horizon strategy, the control variable changes at any new time instant.

12.2.5 Minimum prediction horizon

In many applications, the system under control is characterized by a significant time delay $\theta > 0$ acting at the input or at the output. In the case of input delay, the corresponding model is

$$\begin{aligned} x(k+1) &= Ax(k) + Bu(k-\theta) \\ y(k) &= Cx(k) \end{aligned}$$

In these cases, it is not convenient to penalize in the performance index the future states, outputs, or errors in the time instants k, $k+1$, $k+2$, $k+\theta$, since they do not depend on $u(k+i)$, $i \geq 0$. Rather, the cost function can be modified by defining the minimum prediction horizon $\theta < N_1 < N$, and by considering (system in velocity form and error weighting) the performance index

$$\begin{aligned} J(x(k), u(\cdot), k) &= \sum_{i=N_1}^{N-1} \|y^o(k+i) - y(k+i)\|_Q^2 \\ &+ \sum_{i=0}^{N_u-1} \|\delta u(k+i)\|_R^2 \\ &+ \|y^o(k+N) - y(k+N)\|_S^2 \end{aligned}$$

The use of a minimum prediction horizon is also advisable when the step response of the system exhibits an *inverse response*, due to zeros outside the stability region. In this case, N_1 must be chosen to consider in the cost function only the state, output, or error variables once the inverse response is finished.

12.2.6 Filtering of the reference signal

For constant reference signals, $\bar{y}^o$, it is often suggested to use in the performance index a fictitious reference which at the current time instant k is equal to the controlled variable $y(k)$, and tends to the value $\bar{y}^o$ in the prediction horizon according to a specified law, see the example reported in Figure 12.3.

A possible filtering procedure is given by

$$y^o(k+i) = \left(1 - \alpha^i\right)\bar{y}^o + \alpha^i y(k), \quad 0 \leq \alpha < 1$$

When $\alpha = 0$ there is no filtering, while $\alpha \to 1$ corresponds to slower and slower variations of the fictitious reference signal.
Note that this approach requires to introduce an additional feedback, which can modify the overall stability properties of the resulting closed-loop system.

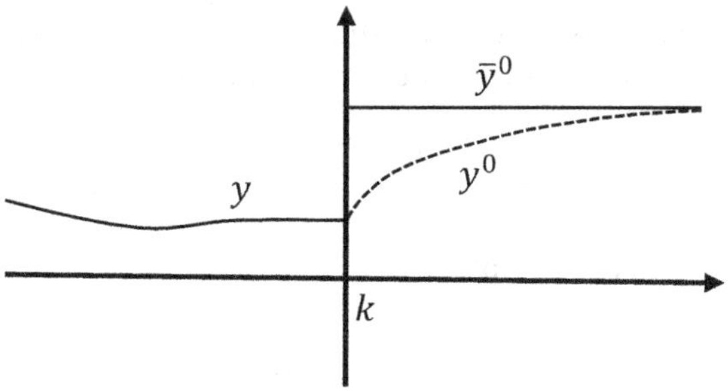

Figure 12.3: Filtering of the reference signal

12.3 *MPC* with impulse or step response models

Asymptotically stable systems can be described by impulse response models characterized by a finite, although possibly large, number of coefficients. These impulse response coefficients can be easily and empirically obtained by means of simple experiments on the plant, so avoiding a long and cumbersome modeling phase.

12.3.1 Impulse response models

In order to describe the main rationale, and simplify the description of these methods, assume that the system under control is $SISO$, asymptotically stable, and described by the state space model

$$\begin{aligned} x(k+1) &= Ax(k) + Bu(k) \\ y(k) &= Cx(k) \end{aligned}$$

The equivalent representation in terms of the impulse response is

$$y(k) = \sum_{i=1}^{\infty} g_i u(k-i)$$

where the impulse response coefficients g_i are

$$g_i = CA^{i-1}B, \quad i > 0$$

In view of the stability assumption, $A^i \overset{i\to\infty}{\to} 0$, so that $g_i \overset{i\to\infty}{\to} 0$. In practice, after M time instants it is possible to assume

$$g_{M+i} = 0, \quad i > 0$$

Note that this corresponds to assume that the system is described by the *Finite Impulse Response* (FIR) model with transfer function

$$G(z) = \frac{g_1 z^{M-1} + ... + g_M}{z^M}$$

Therefore, the model of the plant is

$$y(k) = \sum_{i=1}^{M} g_i u(k-i)$$

and, at time k, the output prediction at time $k+i$, $i > 0$, is

$$y(k+i) = \sum_{j=1}^{i} g_j u(k+i-j) + \sum_{j=i+1}^{M} g_j u(k+i-j) \tag{12.9}$$

For $N < M$,

$$Y(k) = \mathcal{B}_g U(k) + \mathcal{B}_g^{old} U_{old}(k)$$

where

$$\mathcal{B}_g = \begin{bmatrix} g_1 & 0 & 0 & \cdots & 0 & 0 \\ g_2 & g_1 & 0 & \cdots & 0 & 0 \\ \cdots & \cdots & \cdots & \cdots & \cdots & \cdots \\ g_{N-1} & g_{N-2} & g_{N-3} & \cdots & g_1 & 0 \\ g_N & g_{N-1} & g_{N-2} & \cdots & g_2 & g_1 \end{bmatrix}, \quad U_{old}(k) = \begin{bmatrix} u(k-1) \\ u(k-2) \\ \vdots \\ u(k-M+2) \\ u(k-M+1) \end{bmatrix}$$

$$\mathcal{B}_g^{old} = \begin{bmatrix} g_2 & g_3 & \cdots & \cdots & \cdots & g_{M-1} & g_M \\ g_3 & g_4 & \cdots & \cdots & \cdots & g_M & 0 \\ \cdots & \cdots & \cdots & \cdots & \cdots & \cdots & \cdots \\ g_{N+1} & g_{N+2} & \cdots & g_M & 0 & 0 & 0 \end{bmatrix}$$

As in the previous developments, the future outputs depend on the current and future values of the input (vector $U(k)$) and on its past values (vector $U_{old}(k)$). It is then possible to mimic all the previous considerations to formulate and solve the optimization problem.

This approach can be easily extended to systems with multiple inputs (m) and multiple outputs (p); in these cases the coefficients g_i are matrices of dimension (p, m). On the other hand, even in the case of systems with simple dynamics, such as first order systems, it can be necessary to consider a very large number M of coefficients (note however that this depends on the adopted sampling time, when a continuous time system is controlled). Finally, from (12.9), it is apparent that the prediction of $y(k+i)$ is *in open loop*, since it is only based on the past values of the control variable, while it does not depend on the available measurements of the output at time k. This is a very critical drawback, in particular when disturbances act on the system. In order to solve this problem, it is often assumed that a disturbance

$$d(k) = y(k) - \sum_{i=1}^{M} g_i u(k-i)$$

acts on the system, and

$$d(k+i) = d(k), \quad i > 0$$

Letting

$$y(k+i) = \sum_{j=1}^{M} g_j u(k+i-j) + d(k+i)$$

in view of the previous assumptions, it is possible to write

$$y(k+i) = \sum_{j=1}^{M} g_j u(k+i-j) + y(k) - \sum_{j=1}^{M} g_j u(k-j)$$

that is

$$\begin{aligned} y(k+i) &= \left(\sum_{j=1}^{i} g_j u(k+i-j) \right) \\ &+ \left(y(k) + \sum_{j=i+1}^{M} g_j u(k+i-j) - \sum_{j=1}^{M} g_j u(k-j) \right) \end{aligned}$$

As usual, the future output depends on a term which is function of the current and future inputs, and on a second one which is based on the available knowledge at the current time instant k. Therefore, all the previous considerations can be used to formulate and solve an optimization problem where the performance index to be minimized weights the future outputs, errors, and control variables. Finally note that in this approach an observer is not required even when the state of the system is nonmeasurable. This is obviously due to the adopted input-output representation.

12.3.2 Step response models

Now recall that the step response coefficients s_i, $i > 0$, of a *SISO* system can be computed from the impulse response coefficients g_i with the following relation

$$s_i = \sum_{j=1}^{i} g_j$$

or, viceversa,

$$g_i = s_i - s_{i-1}$$

Letting $s_0 = 0$, one obtains

$$y(k) = \sum_{i=1}^{\infty} g_i u(k-i) = \sum_{i=1}^{\infty} (s_i - s_{i-1}) u(k-i) = \sum_{i=1}^{\infty} s_i \delta u(k-i)$$

For asymptotically stable systems, letting $g_{M+i} = 0$, $i > 0$, it follows that

$$s_{M+i} = s_M \tag{12.10}$$

Now, assume again that a disturbance acts on the system output

$$y(k+i) = \sum_{j=1}^{\infty} s_j \delta u(k+i-j) + d(k+i)$$

where

$$d(k+i) = d(k) = y(k) - \sum_{j=1}^{\infty} s_j \delta u(k-j)$$

By recalling (12.10), simple computations lead to

$$y(k+i) = \left(\sum_{j=1}^{i} s_j \delta u(k+i-j) \right) + \left(y(k) + \sum_{j=1}^{M-1} (s_{i+j} - s_j) \delta u(k-j) \right)$$

This expression can be used to compute the output predictions required by the formulation of a predictive control problem, the solution to which is the sequence of future control variations $\delta u(k+i)$, $i = 0, ..., N-1$. Note that, in so doing, an integral action is included into the loop, in fact, the control variable to be applied at any time instant is

$$u(k) = u(k-1) + \delta u^o(k)$$

where $\delta u^o(k)$ is the first element of the optimal control sequence computed at time k.

The most popular algorithm based on the step response coefficients is called *Dynamic Matrix Control* (*DMC*).

Finally note that, also in the case of step response models, an observer is not required for the implementation of the *MPC* control law.

12.4 *MPC* with transfer function models

Assume again that the system is Single-Input, Single-Output and described by the transfer function

$$G(z) = \frac{B(z)}{A(z)} = \frac{b_{n-1}z^{n-1} + b_{n-2}z^{n-2} + ... + b_0}{z^n + a_{n-1}z^{n-1} + a_{n-2}z^{n-2} + ... + a_0}$$

In this case, a simple way to implement a predictive control algorithm is to resort to a minimal canonical form and to implement a state-space formulation of *MPC*.

Alternatively, it is possible to describe the system in the following nonminimal state-space form

$$\begin{aligned} x(k+1) &= Ax(k) + Bu(k) \\ y(k) &= Cx(k) \end{aligned}$$

where

$$x(k+1) = \begin{bmatrix} y(k+1) \\ y(k) \\ \vdots \\ y(k-n+2) \\ u(k) \\ u(k-1) \\ \vdots \\ u(k-n+2) \end{bmatrix}$$

$$A = \begin{bmatrix} -a_{n-1} & -a_{n-2} & \cdots & -a_0 & b_{n-2} & b_{n-3} & \cdots & b_0 \\ 1 & 0 & \cdots & 0 & 0 & 0 & \cdots & 0 \\ \cdots & \cdots & \cdots & \cdots & \cdots & \cdots & \cdots & \cdots \\ 0 & 0 & \cdots & 0 & 0 & 0 & 0 & 0 \\ 0 & 0 & \cdots & 0 & 0 & 0 & \cdots & 0 \\ 0 & 0 & \cdots & 0 & 1 & 0 & \cdots & 0 \\ \cdots & \cdots & \cdots & \cdots & \cdots & \cdots & \cdots & \cdots \\ 0 & 0 & \cdots & 0 & 0 & 0 & 0 & 0 \end{bmatrix}$$

$$B = \begin{bmatrix} b_{n-1} \\ 0 \\ \vdots \\ 0 \\ 1 \\ 0 \\ \vdots \\ 0 \end{bmatrix}, \quad C = \begin{bmatrix} 1 & 0 & \cdots & 0 \end{bmatrix}$$

It is interesting to note that with this state space representation an observer is not required, since the state is known.
Many other algorithms have been developed starting from the knowledge of the transfer function's polynomials $A(z)$ and $B(z)$. Among them, it is worth mentioning the *Generalized Predictive Control* (*GPC*) method, which has been widely used in many different fields.

12.5 Hierarchical structure of the control system

Predictive control algorithms are usually applied at an intermediate level of the overall control structure typical of many industrial processes. At the highest

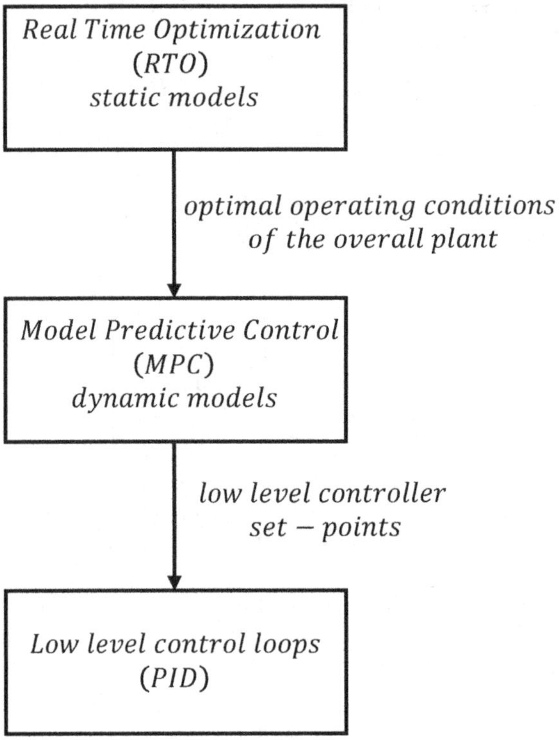

Figure 12.4: A typical industrial hierarchical control scheme.

level, a Real Time Optimization (*RTO*) layer computes the optimal operating conditions of the overall plant, often made by a large number of units (for instance, in the chemical industry, reactors, separators, distillation columns, ...) by means of a very detailed, although static, model of the plant. At the intermediate layer *MPC* dynamically controls the main process variables (temperatures, pressures, concentrations,...) by computing the reference signals for the lower level, where the actuators effectively act at a much higher frequency on the plant manipulated variables. The fast control loops at the lowest layer are usually made by standard *PI-PID* regulators. The corresponding conceptual scheme is shown in Figure 12.4.

12.6 Predictive control of nonlinear systems

Only linear systems have been considered so far. However, the main ideas of *MPC* can also be used to control nonlinear systems of the form

$$x(k+1) = f(x(k), u(k))$$

The performance index to be minimized can be a generic nonlinear function of the state and control variables over the prediction horizon, and the optimization problem can be formulated as

$$\min_{U(k)} \quad J = \sum_{i=0}^{N-1} l(x(k+i), u(k+i))$$

where $l(x, u)$ is a suitable function selected by the designer, and the optimization problem can include state and control constraints.
This problem can be solved numerically, by means of iterative optimization algorithms requiring, at any iteration, to simulate the system. In fact, in the nonlinear case, the prediction of the state along the considered horizon does not depend linearly on the future control moves, so that explicit formulas cannot be obtained even in unconstrained control problems.

12.6.1 Continuous-time *MPC*

In many industrial applications, ***MPC*** is used to control continuous-time nonlinear models of the form

$$\dot{x}(t) = f(x(t), u(t))$$

The control law is assumed to be piecewise continuous and of the form

$$u(t) = u(kT) \quad , \quad t \in [kT, (k+1)T)$$

where T is the adopted sampling period and k is the discrete-time index corresponding to the sampling time kT.
Now, letting

$$U(k) = \begin{bmatrix} u(kT) \\ u(kT+T) \\ \vdots \\ u(kT+NT-T) \end{bmatrix}$$

the optimization problem can be formulated as follows

$$\min_{U(k)} \quad J = \int_{kT}^{kT+NT} l(x(\tau), u(\tau)) d\tau$$

subject to the constraints due to the system's dynamics, and to possible state and control constraints.
Also in this case, only numerical solutions can be computed through iterative optimization algorithms.

12.7 *MPC* algorithms and stability

In the following some formulations of ***MPC*** guaranteeing closed-loop stability are described for linear and nonlinear systems. For simplicity only the state regulation problem is considered and the state is assumed to be measurable.

Let the system under control be described by

$$x(k+1) = f(x(k), u(k)) \tag{12.11}$$

where $f(x, u)$ is continuously differentiable with respect to its arguments, $f(0,0) = 0$, while the state x and the input u must satisfy the constraints

$$x \in X, \quad u \in U \tag{12.12}$$

where X and U are closed and bounded sets including the origin.

For system (12.11) assume to know an *auxiliary control law*

$$u = \kappa_a(x) \tag{12.13}$$

such that $\kappa_a(0) = 0$, and a positively invariant set $X_f \subset X$, including the origin, such that, for the closed-loop system

$$x(k+1) = f(x(k), \kappa_a(x(k)))$$

it holds that $x(\bar{k}) \in X_f$ implies

$$\begin{aligned} x(k) &\in X_f, \quad k \geq \bar{k} \\ u(k) &= \kappa_a(x(k)) \in U, \quad k \geq \bar{k} \end{aligned}$$

In other words, given an initial state $x(\bar{k})$ in X_f, the evolution of the state of the closed-loop system with the auxiliary control law remains in X_f and the state and input constraints (12.12) are satisfied.

Consider now the problem of computing at time k the control sequence $u(k)$, $u(k+1)$, ..., $u(k+N-1)$ that minimizes the finite horizon cost function

$$J(x(k), u(\cdot), k) = \sum_{i=0}^{N-1} \left(\|x(k+i)\|_Q^2 + \|u(k+i)\|_R^2 \right) + V_f(x(k+N)) \tag{12.14}$$

subject to the dynamic constraint (12.11) and

$$\begin{aligned} x(k+i) &\in X, \quad u(k+i) \in U \\ x(k+N) &\in X_f \end{aligned}$$

In the cost function (12.14) V_f is a terminal penalty such that $V_f(x) \geq 0$, $V_f(0) = 0$, and $Q = Q' > 0$, $R = R' > 0$ are matrices of suitable dimension. Note that for simplicity the matrix Q is assumed to be strictly positive; however also $Q \geq 0$ could be used provided that a suitable observability assumption is verified.

From the solution to the finite horizon optimization problem, and by means of the receding horizon principle, the implicit state-feedback time-invariant control law is derived

$$u = \kappa^{MPC}(x) \tag{12.15}$$

Let $J^o(x(k), k)$ be the optimal value of the cost function and assume that J^o is a continuous function of its arguments. Then, under the previous assumptions, the following results can be proven.

Theorem 12.7.1. *If, for any $x \in X_f$,*

$$V_f\left(f\left(x(k), \kappa_a(x(k))\right)\right) - V_f(x(k)) + \left(\|x(k)\|_Q^2 + \|\kappa_a(x(k))\|_R^2\right) \leq 0 \qquad (12.16)$$

then the origin $x = 0$ is an asymptotically stable equilibrium point of the closed-loop system obtained with the MPC control law (12.15).

Proof. Let the solution to the optimization problem, at the generic time instant k, be given by

$$U^o(k) = \left[\begin{array}{cccccc} u_k^o(k) & u_k^o(k+1) & \cdots & u_k^o(k+N-2) & u_k^o(k+N-1) \end{array}\right]$$

and define by $J^o(x(k), k)$ the associated cost function. Note that $J^o(x(k), k) \geq 0$ in view of the assumptions on Q, R, and V_f. According to the RH principle, apply the first control action $u_k^o(k)$, then, at the next time instant $k+1$, the state is

$$x^o(k+1) = f\left(x(k), u_k^o(k)\right)$$

By applying the remaining part of the optimal sequence $U^o(k)$, one would obtain $x^o(k+N) \in X_f$, so that, at time $k+1$, the control sequence

$$\tilde{U}(k+1) = \left[\begin{array}{ccccc} u_k^o(k+1) & u_k^o(k+2) & \cdots & u_k^o(k+N-1) & \kappa_a(x(k+N)) \end{array}\right]$$

is a feasible one for the optimization problem stated at $k+1$. However, the cost function $\tilde{J}(x^o(k+1), k+1)$ corresponding to $\tilde{U}(k+1)$ is not a priori optimal for the problem formulated at time $k+1$, so that

$$J^o(x^o(k+1), k+1) \leq \tilde{J}(x^o(k+1), k+1) \qquad (12.17)$$

From (12.14), by considering all the terms of J, it follows that

$$\begin{aligned} & \tilde{J}(x^o(k+1), k+1) - J^o(x(k), k) \\ = & -\left(\|x(k)\|_Q^2 + \left\|\kappa^{MPC}(x(k))\right\|_R^2\right) \\ & + V_f\left(f\left(x^o(k+N), \kappa_a(x^o(k+N))\right)\right) - V_f(x^o(k+N)) \\ & + \left(\|x^o(k+N)\|_Q^2 + \|\kappa_a(x^o(k+N))\|_R^2\right) \end{aligned}$$

Therefore, by also recalling (12.17), the sufficient condition (12.16) guarantees that

$$J^o(x^o(k+1), k+1) - J^o(x(k), k) \leq -\left(\|x(k)\|_Q^2 + \left\|\kappa^{MPC}(x(k))\right\|_R^2\right) \leq 0$$

where the equality holds only for $x(k) = 0$, since $Q > 0$. Hence $J^o(x(k), k)$ is a Lyapunov function for the closed-loop system with the *MPC* control law.

Before introducing some specific algorithms for the choice of the terminal penalty V_f and of the terminal constraint X_f satisfying condition (12.16), some remarks are in order.

Remark. The previous stability result is based on the proof that the optimal cost function is decreasing. In order to guarantee this property, note that it is not necessary to determine at any time instant k the optimal (and feasible) sequence $U^o(k)$ that minimizes (12.14), but it is sufficient to compute a sequence

$$\bar{U}(k) = \left[\begin{array}{ccccc} \bar{u}_k(k) & \bar{u}_k(k+1) & \cdots & \bar{u}_k(k+N-2) & \bar{u}_k(k+N-1) \end{array} \right]$$

such that $\bar{J}(x(k), k) \leq \tilde{J}(x(k), k)$ where $\bar{J}(x(k), k)$ and $\tilde{J}(x(k), k)$ are the values of the cost function given the sequences $\bar{U}(k)$ and

$$\tilde{U}(k) = \left[\begin{array}{ccccc} u^o_{k-1}(k) & u^o_{k-1}(k+1) & \cdots & u^o_{k-1}(k+N-2) & \kappa_a(x(k+N-1)) \end{array} \right]$$

This is very important from a computational point of view, since it shows that it is not necessary to compute the optimal solution at any time. Rather, it is required to determine a solution improving the one given by $\tilde{U}(k)$.

Remark. The use of the *MPC* control law (12.15) allows one to enlarge the stability region of the auxiliary control law and improves its performance. In fact, regardless of the choice of V_f and X_f, and provided that the conditions of the previous theorem are satisfied, denoting by $X^{MPC}(N)$ the maximum positively invariant set associated with the closed-loop system with control law (12.15) and prediction horizon N, where $x \in X$ and $\kappa^{MPC}(x) \in U$, the following results hold:

1. $X^{MPC}(N+1) \supseteq X^{MPC}(N)$, that is the size of X^{MPC} increases with N;
2. $X^{MPC}(N) \supseteq X_f$, since the auxiliary control law $u = \kappa_a(x)$ is a feasible solution for the optimization problem;
3. there exists a value $\bar{N}$ such that $X^{MPC}(\bar{N}) \supseteq \bar{X}_f$, where $\bar{X}_f$ is the maximum positively invariant set, a priori unknown, associated with the auxiliary control law (12.13).

Two algorithms that satisfy the assumptions required by the previous stability theorem are now introduced.

12.7.1 Zero terminal constraint

The method is characterized by the following choices

$$\kappa_a(x) = 0 \tag{12.18}$$
$$X_f = \{0\} \tag{12.19}$$
$$V_f = 0 \tag{12.20}$$

i.e. the state at the end of the finite horizon is required to be at the origin.

Since by assumption $f(0,0) = 0$, X_f is a positive invariant set for the closed-loop system with the auxiliary control law (12.18) and $\kappa_a(x)$ satisfies also the input constraints. Moreover, since the terminal penalty is null (see (12.20)), the condition (12.16) is satisfied (with the equal) so that, in view of Theorem 12.7.1, the asymptotically stability of the origin is proven.

This approach, although conceptually very simple, is interesting mainly for linear systems, where satisfying the terminal equality constraint (12.19) is simpler. However note that, when input constraints are included, the set $X^{RH}(N)$ coincides with the set $X^{con}(N)$, defined as the set of states that can be steered to the origin in N steps with a bounded control. This set $X^{con}(N)$ can be very small in many cases, consider for example an unstable linear system to be controlled with a saturated input variable.
More in general, if the system under control is nonlinear, fulfilling the terminal equality constraint can be numerically very difficult in a finite number of iterations, even without any limitation on the input variable, so that other approaches are required.

12.7.2 Quasi infinite horizon *MPC*

For simplicity, the method will be first presented for linear systems, then it will be extended to the nonlinear case.

Linear systems

Assume that the system is described by

$$x(k+1) = Ax(k) + Bu(k)$$

For this system compute the infinite horizon LQ optimal control law

$$u(k) = -K_{LQ}x(k) \tag{12.21}$$

with the penalty matrices Q and R used also in the ***MPC*** cost function, and let P be the positive definite solution to the associated Riccati equation

$$(A - BK_{LQ})' P (A - BK_{LQ}) - P = -\left(Q + K'_{LQ}RK_{LQ}\right)$$

Consider the LQ control law (12.21) as the auxiliary control law, define the terminal inequality constraint as

$$X_f = \{x | x'Px \leq \alpha\} \subset X$$

where α is a sufficient small positive real number, and choose the terminal penalty as

$$V_f(x) = x'Px \tag{12.22}$$

These choices fulfill the sufficient condition (12.16). In fact

$$\begin{aligned}
& \Gamma(x(k)) \\
\triangleq\ & V_f\left(f\left(x(k), -K_{LQ}x(k)\right)\right) - V_f\left(x(k)\right) + \left(\|x(k)\|_Q^2 + \|K_{LQ}x(k)\|_R^2\right) \\
=\ & x'(k)\left\{(A - BK_{LQ})' P (A - BK_{LQ}) - P + \left(Q + K'_{LQ}RK_{LQ}\right)\right\} x(k) \\
=\ & 0
\end{aligned}$$

Moreover X_f is positively invariant for the closed-loop with the auxiliary control law (12.21), since it is bounded by a level line of the Lyapunov function associated to the corresponding closed-loop system. Finally, in a neighbourhood of the origin it is also true that $u = -K_{LQ}x \in U$.

Note that the terminal penalty (12.22) represents the infinite horizon cost from $k + N$ to infinity in a classical unconstrained LQ problem.

Nonlinear systems

First consider the expansion of the nonlinear model at the origin

$$\begin{aligned} f(x,u) &= \left.\frac{\partial f}{\partial x}\right|_{x=u=0} \delta x + \left.\frac{\partial f}{\partial u}\right|_{x=u=0} \delta u + \phi(\delta x, \delta u) \\ &= A\delta x + B\delta u + \phi(\delta x, \delta u) \end{aligned}$$

where

$$\lim_{\|(\delta x, \delta u)\| \to 0} \sup \frac{\|\phi(\delta x, \delta u)\|}{\|(\delta x, \delta u)\|} = 0$$

With the same penalty matrices Q and R of the stage cost of the *MPC* optimization problem, compute the infinity LQ control law

$$\delta u(k) = -K_{LQ}\delta x(k)$$

for the linearized system described by the matrices A and B. Then, it follows that

$$f(x,u) = (A - BK_{LQ})\delta x + \phi(\delta x, -K_{LQ}\delta x)$$

and

$$\lim_{\|\delta x\| \to 0} \sup \frac{\|\phi(\delta x, -K_{LQ}\delta x)\|}{\|(\delta x, K_{LQ}\delta x)\|} = 0$$

Now, solve the Lyapunov equation

$$(A - BK_{LQ})' \Pi (A - BK_{LQ}) - \Pi = -\beta \left(Q + K'_{LQ}RK_{LQ}\right)$$

with $\beta > 1$.

Finally consider the auxiliary control law

$$u(k) = -K_{LQ}x(k) \tag{12.23}$$

and the terminal penalty

$$V_f(x) = x'\Pi x \tag{12.24}$$

With this choices, the sufficient condition (12.16) is verified. In fact, one has

$$\begin{aligned} &\Gamma(x(k)) \\ \triangleq\ & V_f\left(f\left(x(k), -K_{LQ}x(k)\right)\right) - V_f\left(x(k)\right) + \left(\|x(k)\|_Q^2 + \|K_{LQ}x(k)\|_R^2\right) \\ =\ & f\left(x(k), -K_{LQ}x(k)\right)' \Pi f\left(x(k), -K_{LQ}x(k)\right) \\ & -x'(k)\Pi x(k) + x'(k)Qx(k) + x'(k)K'_{LQ}RK_{LQ}x(k) \end{aligned}$$

or, in the neighbourhood of the origin and defining $\bar{\phi}(x) = \phi(x, -K_{LQ}x)$,

$$\begin{aligned}
&\Gamma(x(k)) \\
= \; & x'(k)\,(A - BK_{LQ})'\,\Pi\,(A - BK_{LQ})\,x(k) + 2x'(k)(A - BK_{LQ})'\Pi\bar{\phi}(x(k)) \\
& +\bar{\phi}'(x)\Pi\bar{\phi}(x) + x'(k)\left(Q + K'_{LQ}RK_{LQ}\right)x(k) - x(k)'\Pi x(k) \\
= \; & x'(k)\left\{(A - BK_{LQ})'\,\Pi\,(A - BK_{LQ}) - \Pi + \left(Q + K'_{LQ}RK_{LQ}\right)\right\}x(k) \\
& +2x'(k)(A - BK_{LQ})'\Pi\bar{\phi}(x(k)) + \bar{\phi}'(x)\Pi\bar{\phi}(x) \\
= \; & x'(k)\,(1 - \beta)\left(Q + K'_{LQ}RK_{LQ}\right)x(k) \\
& +2x'(k)(A - BK_{LQ})'\Pi\bar{\phi}(x(k)) + \bar{\phi}'(x)\Pi\bar{\phi}(x)
\end{aligned}$$

Now note that, for any constant $L_{\bar{\phi}} > 0$, it is possible to find a neighbourhood of the origin such that

$$2x'(k)(A - BK_{LQ})'\Pi\bar{\phi}(x(k)) \leq 2\,\|\Pi\|\,\|A - BK_{LQ}\|\,L_{\bar{\phi}}\,\|x(k)\|^2$$

and

$$\bar{\phi}'(x)\Pi\bar{\phi}(x) \leq \|\Pi\|\,L^2_{\bar{\phi}}\,\|x(k)\|^2$$

It follows that there exist a constant $L_{\bar{\phi}} > 0$ such that $\Gamma(x(k)) \leq 0$ in a sufficiently small neighbourhood of the origin, and the sufficient condition (12.16) is satisfied.

Finally considering the set

$$X_f = \{x|\, x'\Pi x \leq \alpha\} \subset X$$

it is apparent that X_f is positively invariant for the closed-loop system with the auxiliary control law (12.23), since it is limited by a level line of a Lyapunov function associated to the closed-loop linearized system. Moreover, as already proven, for a sufficiently small α (12.16) is verified and by continuity $u = -K_{LQ}x \in U$.

A
Matrix theory and norms

A.1 Norms

The norm of e is now defined, where e can be

- a vector
- a matrix
- a time-based signal
- a system, defined by the matrix $G(s)$ in the frequency, or by $g(t)$ function of time

Definition A.1.1. A norm of e is a real number, represented by the symbol $\|e\|$, satisfying the following properties:

1. non negativity: $\|e\| \geq 0$
2. positivity: $\|e\| = 0 \Leftrightarrow e = 0$
3. homogeneity: $\|\alpha e\| = |\alpha| \, \|e\|$ for any real or complex scalar α
4. triangular inequality: $\|e_1 + e_2\| \leq \|e_1\| + \|e_2\|$

A.1.1 Norms of vectors

Definition A.1.2. Norm p

$$\|e\|_p = \left(\sum_{i=1}^{m} |e_i|^p \right)^{1/p}$$

Definition A.1.3. 2-norm (Euclidean norm, or minimal distance between two points)

$$\|e\|_2 = \sqrt{\sum_{i=1}^{m} e_i^2}$$

Definition A.1.4. ∞ -norm

$$\|e\|_\infty = \sup_i |e_i|$$

A.1.2 Norms of matrices

Definition A.1.5. The norm of a matrix, $\|A\|$, must satisfy the multiplicative property

$$\|AB\| \leq \|A\| \, \|B\|$$

Definition A.1.6. The induced norm p is

$$\|A\|_{ip} = \sup_{d \neq 0} \frac{\|Ad\|_p}{\|d\|_p}$$

Definition A.1.7. The induced 2-norm is

$$\|A\|_{i2} = \sup_{d \neq 0} \frac{\|Ad\|_2}{\|d\|_2} = \bar{\sigma}(A)$$

A.1.3 Norms of signals

Definition A.1.8. The p-norm is

$$\|e(t)\|_p = \left(\int_{-\infty}^{\infty} \sum_{i=1}^{m} |e_i(\tau)|^p \, d\tau \right)^{1/p}$$

Definition A.1.9. The 2-norm is

$$\|e(t)\|_2 = \left(\int_{-\infty}^{\infty} \sum_{i=1}^{m} |e_i(\tau)|^2 \, d\tau \right)^{1/2}$$

Definition A.1.10. The ∞-norm is

$$\|e(t)\|_\infty = \sup_{\tau} \left(\sup_{i} |e_i(\tau)| \right)$$

A.1.4 Norms of systems

Definition A.1.11. The $\mathcal{H}_2$ norm of a strictly proper, asymptotically stable transfer function $G(S)$ is

$$\|G(s)\|_2 = \sqrt{\left\{ \frac{1}{2\pi} \int_{-\infty}^{+\infty} tr \left[G(j\omega) G'(-j\omega) \right] d\omega \right\}}$$

Definition A.1.12. The $\mathcal{H}_\infty$ norm of a proper, asymptotically stable transfer function $G(S)$ is

$$\|G(s)\|_\infty = \sup_{\omega} \bar{\sigma}(G(j\omega))$$

Properties

From the Parseval theorem

$$\|G(s)\|_2 = \|g(t)\|_2 = \sqrt{\left\{\int_0^{+\infty} tr\left[g'(\tau)g(\tau)\right] d\tau\right\}}$$

where $g(t)$ is the impulse response.

The norm $\mathcal{H}_\infty$ coincides with the 2-induced norm in the time domain

$$\|G(s)\|_\infty = \sup_{d(t)\neq 0} \frac{\|y(t)\|_2}{\|d(t)\|_2} = \sup_{\|d(t)\|_2=1} \|y(t)\|_2$$

A.2 Singular values

Definition A.2.1. Unitary matrix

Matrix U (with complex elements) is unitary if $U^* = U^{-1}$ where U^* is the conjugate transpose of U.

The following properties hold

- $|\lambda_i(U)| = 1, \forall i$;
- $\sigma_i(U) = 1, \forall i$ (see below).

Definition A.2.2. The *singular values* of the matrix $\Phi \in C^{p,m}$ are the $k = min(p,m)$ largest roots of the eigenvalues of $\Phi^*\Phi$ or of $\Phi\Phi^*$

$$\begin{aligned}
\sigma_i(\Phi) &: = \sqrt{\lambda_i(\Phi^*\Phi)} = \sqrt{\lambda_i(\Phi\Phi^*)}, \quad m = p \\
\sigma_i(\Phi) &: = \sqrt{\lambda_i(\Phi^*\Phi)}, \quad m > p \\
\sigma_i(\Phi) &: = \sqrt{\lambda_i(\Phi\Phi^*)}, \quad m < p
\end{aligned}$$

Definition A.2.3. Any matrix $\Phi \in C^{p,m}$ can be partitioned with the singular value decomposition

$$\Phi = U\Sigma V^*$$

where the matrices $U \in C^{p\times p}$ and $V \in C^{m\times m}$ are unitary, while the matrix Σ is defined by

$$\begin{aligned}
\Sigma &= \begin{bmatrix} \Sigma_1 \\ 0 \end{bmatrix} \in R^{p\times m}, \quad p \geq m \\
\Sigma &= \begin{bmatrix} \Sigma_1 & 0 \end{bmatrix} \in R^{p\times m}, \quad p \leq m
\end{aligned}$$

where

$$\Sigma_1 = diag\left\{\sigma_1, \sigma_2, \ldots, \sigma_k\right\}; \quad k = \min(p,m)$$

with

$$\sigma_1 \geq \sigma_2 \geq \ldots \geq \sigma_k$$

Moreover, define the maximum singular value $\bar{\sigma} \equiv \sigma_1$ and the minimum singular value

$$\underline{\sigma} \equiv \left\{ \begin{array}{lll} \sigma_m & if & p \geq m \\ 0 & if & p < m \end{array} \right\}$$

U and V are an orthonormal basis of the inputs and outputs spaces.

Definition A.2.4 (Singular values of 2x2 matrices). Let

$$b = tr(\Phi^* \Phi) = \sum_{i,j} |\varphi_{ij}|^2, \quad c = \det(\Phi^* \Phi) = \det(\Phi)^2$$

Therefore, the sum of the eigenvalues of $\Phi^* \Phi$ is $b = \lambda_1 + \lambda_2$, while the product is $c = \lambda_1 \lambda_2$. Since

$$\sigma_i(\Phi) := \sqrt{\lambda_i(\Phi^* \Phi)}$$

it holds that

$$\bar{\sigma}(\Phi) = \sqrt{\frac{b + \sqrt{b^2 - 4c}}{2}}; \quad \underline{\sigma}(\Phi) = \sqrt{\frac{b - \sqrt{b^2 - 4c}}{2}}$$

Definition A.2.5 (Inputs and outputs directions). Let

$$\Phi = U \Sigma V^*$$

The columns of U, denoted by u_i, define the outputs directions, are orthogonal and with unitary length

$$\begin{array}{rcl} \|u_i\|_2 & = & \sqrt{|u_{i1}|^2 + |u_{i2}|^2 + \cdots + |u_{il}|^2} = 1 \\ u_i' u_i & = & 1, \quad u_i' u_j = 0, \quad i \neq j \end{array}$$

The columns V, denoted by v_i, define the inputs directions, are orthogonal and with unitary length.

In fact, $V^* V = I$

$$\Phi V = U \Sigma$$

and for the i-th column

$$\Phi v_i = \sigma_i u_i$$

This means that, if one considers an input with direction specified by v_i, the output direction is u_i. Moreover, since u_i and v_i have unitary norm, one has

$$\|\Phi v_i\|_2 = \|\sigma_i u_i\|_2 = \sigma_i = \frac{\|\Phi v_i\|_2}{\|v_i\|_2}$$

Hence, given the input v_i, the system has gain σ_i and output with direction u_i.

Properties

- The maximum gain corresponds to the maximum singular value of Φ

$$\sup_{w\neq 0}\frac{\|\Phi w\|_2}{\|w\|_2}=\sup_{\|w\|_2=1}\|\Phi w\|_2=\bar{\sigma}(\Phi)$$

- The minimum value of the gain is given by the minimum singular value of Φ

$$\inf_{w\neq 0}\frac{\|\Phi w\|_2}{\|w\|_2}=\inf_{\|w\|_2=1}\|\Phi w\|_2=\underline{\sigma}(\Phi)$$

- For any vector w, it holds that

$$\underline{\sigma}(\Phi)\leq\frac{\|\Phi w\|_2}{\|w\|_2}\leq\bar{\sigma}(\Phi)$$

- If Φ is square and invertible

$$\bar{\sigma}(\Phi^{-1})=\frac{1}{\underline{\sigma}(\Phi)}$$

$$\underline{\sigma}(I+\Phi)\leq\underline{\sigma}(\Phi)+1$$

A.3 Matrix inversion lemma

Given the matrices A, B, C, D of suitable dimensions with A and C non singular

$$(A+BCD)^{-1}=A^{-1}-A^{-1}B\left(DA^{-1}B+C^{-1}\right)^{-1}DA^{-1}$$

Bibliography

The analysis and control of nonlinear systems are treated in:

- Khalil H.K.: "Nonlinear Systems", Prentice Hall, Englewood Cliffs NJ, 1991.
- Sastry S.: "Nonlinear Systems", Springer, 1999.
- Slotine J.J.E., Li W.: "Applied Nonlinear Control". Prentice Hall, Englewood Cliffs NJ, 1991.

The analysis and design of multivariable control systems are extensively discusses in:

- Kailath T.: "Linear systems". Prentice Hall, 1980.
- Maciejowski J.M.: "Multivariable Feedback Design", Addison-Wesley, 1989.
- Glad T., Ljung L.: "Control Theory - Multivariable and Nonlinear Methods", Taylor and Francis, 2000.
- Skogestad S., Postlethwaite I.: "Multivariable Feedback Control, Analysis and Design", Wiley, 2005.
- Zhou K., Doyle J.C., Glover K.: "Robust and Optimal Control", Prentice Hall, 1996.

Fundamental books on optimal control and filtering are:

- Kwakernaak H., Sivan R.: "Linear Optimal Control Systems", http://www.ieeecss.org/PAB/classics/
- Anderson B.D.O., Moore J.B.: "Optimal Control - Linear Quadratic Methods", Prentice Hall, 1989.
- Sepulchre R., Jankovic M., Kokotovic P.: "Constructive Nonlinear Control", Springer Verlag, 1997.
- Anderson B.D.O., Moore J.B.: "Optimal Filtering", Dover Publications, 2005.

Model Predictive Control is presented in:

- Camacho E.F., Bordons C.: "Model Predictive Control in the Process Industry", Springer Verlag, 1995.
- Maciejowski J.M.: "Predictive Control with Constraints", Prentice Hall, UK, 2001.
- Rawlings J.B., Mayne D.Q.: "Model Predictive Control: Theory and Design", Nob Hill Publishing, USA, 2009.